The Dappled World

A Study of the Boundaries of Science

It is often supposed that the spectacular successes of our modern mathematical sciences support a lofty vision of a world completely ordered by one elegant theory. In this book Nancy Cartwright argues to the contrary. When we draw our image of the world from the way modern science works – as empiricism teaches us we should – we end up with a world where some features are precisely ordered, others are given to rough regularity and still others behave in their own diverse ways. This patchwork of laws makes sense when we realise that laws are very special productions of nature, requiring very special arrangements for their generation. Combining previously published and newly written essays on physics and economics, *The Dappled World* carries important philosophical consequences and offers serious lessons for both the natural and the social sciences.

Nancy Cartwright is Professor of Philosophy at the London School of Economics and Political Science and at the University of California, San Diego, a Fellow of the British Academy, and a MacArthur Fellow. She is the author of *How the Laws of Physics Lie* (1983), *Nature's Capacities and their Measurement* (1989), and *Otto Neurath: Philosophy Between Science and Politics*, co-authored with Jordi Cat, Lola Fleck and Thomas Uebel (1995).

To Emily and Sophie

Contents

Acknowledgements

This book is squarely in the tradition of the Stanford School and is deeply influenced by the philosophers of science I worked with there. It began with the pragmatism of Patrick Suppes and the kinds of views he articulated in his *Probabilistic Metaphysics*.[1] Then there was Ian Hacking, John Dupré, Peter Galison and, for one year, Margaret Morrison.

The second major influence is the Modelling and Measurement in Physics and Economics research group at the London School of Economics. The Modelling portion of the project was directed by Mary Morgan and Margaret Morrison; Measurement, by Mary Morgan and Hasok Chang. I have been helped by the ideas and the studies of all of the research assistants in the project, whose own work on the topics has been a model I would wish to follow for philosophical originality and getting the details right: Towfic Shomar, Mauricio Suárez, Marco Del Seta, Cynthia Ma, George Zouros, Antigone Nounou, Francesco Guala, Sang Wook Yi, Julian Reiss and Makiko Ito. I have worked out almost all of the ideas in these chapters in detailed conversations and debates with Jordi Cat.

Julian Reiss and Sang Wook Yi have read and criticised the entire book, which has improved as a result. The production of the typescript was impressively carried out by Dorota Rejman, with the help of Julian Reiss and Sang Wook Yi. The original drawings are by Rachel Hacking; the machines by Towfic Shomar.

I am very grateful to the MacArthur Foundation and the LSE Centre for Philosophy of Natural and Social Science for financial support throughout, and to the Latsis Foundation for a grant that allowed me to complete the book.

Much of the material of this book has been drawn from articles published elsewhere. Much has not been published before. The exact origin of each chapter is described in the acknowledgements section at the end of it.

[1] Suppes 1984.

Introduction

This book supposes that, as appearances suggest, we live in a dappled world, a world rich in different things, with different natures, behaving in different ways. The laws that describe this world are a patchwork, not a pyramid. They do not take after the simple, elegant and abstract structure of a system of axioms and theorems. Rather they look like – and steadfastly stick to looking like – science as we know it: apportioned into disciplines, apparently arbitrarily grown up; governing different sets of properties at different levels of abstraction; pockets of great precision; large parcels of qualitative maxims resisting precise formulation; erratic overlaps; here and there, once in a while, corners that line up, but mostly ragged edges; and always the cover of law just loosely attached to the jumbled world of material things. For all we know, most of what occurs in nature occurs by hap, subject to no law at all. What happens is more like an outcome of negotiation between domains than the logical consequence of a system of order. The dappled world is what, for the most part, comes naturally: regimented behaviour results from good engineering.

I shall focus on physics and economics, for these are both disciplines with imperialist tendencies: they repeatedly aspire to account for almost everything, the first in the natural world, the second in the social. Since at least the time of the Mechanical Philosophy, physicists have been busy at work on a theory of everything. For its part, contemporary economics provides models not just for the prices of the rights for off-shore oil drilling, where the market meets very nice conditions, but also for the effects of liberal abortion policies on teenage pregnancies, for whom we marry and when we divorce and for the rationale of political lobbies.

My belief in the dappled world is based in large part on the failures of these two disciplines to succeed in these aspirations. The disorder of nature is apparent. We need good arguments to back the universal rule of law. The successes of physics and the 'self evidence' of the governing principles in economics – the assumption that we act to maximise our own utilities, for example, or the fact that certain behaviours can be *proved* to be rational by game theory – are supposed to supply these arguments. I think they show just the opposite. They show a world whose laws are plotted and pieced.

Consider physics first. I look particularly at quantum physics, because it is what many suppose – in some one or another of its various guises – to be the governor of all of matter.[2] I also look to some extent at classical physics, both classical mechanics and classical electromagnetic theory. For these have an even more firmly established claim to rule, though the bounds of their empire have contracted significantly since the pretensions of the seventeenth century Mechanical Philosophy or the hopes for an electromagnetic take-over at the end of the nineteenth century. And I look at the relations between them. Or rather, I look at a small handful of cases out of a vast array, and perhaps these are not even typical, for the relations among these theories are various and complicated and do not seem to fit any simple formulae.

The conventional story of scientific progress tells us that quantum physics has replaced classical physics. We have discovered that classical physics is false and quantum physics is, if not true, a far better approximation to the truth. But we all know that quantum physics has in no way replaced classical physics. We use both; which of the two we choose from one occasion to another depends on the kinds of problems we are trying to solve and the kinds of techniques we are master of. 'Ah', we are told, 'that is only in practice. In principle everything we do in classical physics could be done, and done more accurately, in quantum physics.' But I am an empiricist. I know no guide to principle except successful practice. And my studies of the most successful applications of quantum theory teach me that quantum physics works in only very specific kinds of situations that fit the very restricted set of models it can provide; and it has never performed at all well where classical physics works best.

This is how I have come to believe in the patchwork of law. Physics in its various branches works in pockets, primarily inside walls: the walls of a laboratory or the casing of a common battery or deep in a large thermos, walls within which the conditions can be arranged *just so*, to fit the well-confirmed and well-established models of the theory, that is, the models that have proved to be dependable and can be relied on to stay that way. Very

[2] Looked at from a different point of view it is superstring theory that makes the loudest claims right now to be a theory of everything. But superstring theory is not (yet?) a theory of the physical world, it is a speculation; and even its strongest advocates do not credit it with its own empirical content. Mathematics, they say in its defence, is the new laboratory site for physics. (See Galison forthcoming for a discussion of this.) Its claims to account for anything at all in the empirical world thus depend on the more pedestrian theories that it purports to be able to subsume; that is, a kind of 'trickle down' theory of universal governance must be assumed here. So, setting aside its own enormous internal problems, the empire of superstring theory hangs or falls with those of all the more long standing theories of physics that do provide detailed accounts of what happens in the world.

occasionally it works outside of walls as well, but these are always in my investigations cases where nature fortuitously resembles one of our special models without the enormous design and labour we must normally devote to making it do so.

Carl Menger thought that economics works in the same way.[3] Menger is one of the three economists credited with the notion of marginal utility. He is also famous for his attack on historical economics and his insistence that economics should be a proper science. By 'proper' science he meant one that uses precise concepts which have exact deductive relations among them. His paradigm was $\mathbf{F} = \boldsymbol{ma}$. In mechanics we do get this kind of exact relation, but at the cost of introducing abstract concepts like *force*, concepts whose relation to the world must be mediated by more concrete ones. These more concrete concepts, it turns out, are very specific in their form: the forms are given by the *interpretative models* of the theory, for example, two compact masses separated by a distance r, the linear harmonic oscillator, or the model for a charge moving in a uniform magnetic field. This ensures that 'force' has a very precise content. But it also means that it is severely limited in its range of application. For it can be attached to only those situations that can be represented by these highly specialised models. This is just what Menger said we should expect from economics: we can have concepts with exact deductive relations among them but those concepts will not be ones that readily represent arrangements found in 'full empirical reality'. This does not mean they never occur, but if they do it will be in very special circumstances.

Much of the theorising we do in economics right now goes in exactly the opposite direction to that recommended by Menger. But in the end this is no aid to any aspirations economics might have to take over the entire study of social and economic life. Economics and physics are both limited in where they govern, though the limitations have different origins. Contemporary economics uses not abstract or theoretical or newly invented concepts – like 'force' or 'energy' or 'electromagnetic field' – concepts partially defined by their deductive relations to other concepts, but rather very mundane concepts that are unmediated in their attachment to full empirical reality.[4] We study, for instance, the capacity of skill loss during unemployment to produce persistence in employment shocks, or the limited enforceability of debt contracts, or whether the effect of food shortages on famines is mediated by failures of food entitlement arrangements. Nevertheless we want our treatments to be rigorous and our conclusions to follow deductively. And the way you get deductivity when you do not have it in the concepts is to put enough of the right kind of structure into the model. That is the trick of building a model

[3] Menger 1883 [1963].
[4] *Cf.* Mäki 1996.

in contemporary economics: you have to figure out some circumstances that are constrained in just the right way that results can be derived deductively.

My point is that theories in physics and economics get into similar situations but by adopting opposite strategies. In both cases we can derive consequences rigorously only in highly stylised models. But in the case of physics that is because we are working with abstract concepts that have considerable deductive power but whose application is limited by the range of the concrete models that tie its abstract concepts to the world. In economics, by contrast, the concepts have a wide range of application but we can get deductive results only by locating them in special models.

Specifically, I shall defend three theses in this book:

(1) The impressive empirical successes of our best physics theories may argue for the truth of these theories but not for their universality. Indeed, the contrary is the case. The very way in which physics is used to generate precise predictions shows what its limits are. The abstract theoretical concepts of high physics describe the world only via the models that interpret these concepts more concretely. So the laws of physics apply only where its models fit, and that, apparently, includes only a very limited range of circumstances. Economics too, though for almost opposite reasons, is confined to those very special situations that its models can represent, whether by good fortune or by good management.

(2) Laws, where they do apply, hold only *ceteris paribus*. By 'laws' I mean descriptions of what regularly happens, whether regular associations or singular causings that occur with regularity, where we may, if we wish, allow counterfactual as well as actual regularities or add the proviso that the regularities in question must occur 'by necessity'. Laws hold as a consequence of the repeated, successful operation of what, I shall argue, is reasonably thought of as a *nomological machine*.

(3) Our most wide-ranging scientific knowledge is not knowledge of laws but knowledge of the *natures* of things, knowledge that allows us to build new nomological machines never before seen giving rise to new laws never before dreamt of.

In considering my defence of these three theses it will help to understand the motives from which I approach the philosophy of science. I notice among my colleagues three very different impulses for studying science. It is a difference that comes out when we ask the question: why are even deeply philosophical historians of physics generally not interested in leading philosophers of physics and vice versa? Or what is the difference between the philosopher of biology John Dupré, with whom I have so much in common, and philosophers studying the mathematical structures of our most modern theories in physics, from whom I differ so radically despite our shared interest in physics? These latter, I would say, are primarily interested in the world that science represents. They are interested, for instance, in the geometry of space and time. Their interest in science generally comes from their belief that understanding our most advanced scientific representations of the world is their best route to understanding that world itself. John Dupré too is interested in the world, but in the material, cultural and politico-economic world of day-to-day and historical life.[5] He is interested in science as it affects that life, in all the ways that it affects that life. Hence he is particularly interested in the politics of science, not primarily the little politics of laboratory life that shapes the internal details of the science but its big politics that builds bombs and human genomes. That kind of interest is different again from most historians and sociologists of science whose immediate object is science itself, but – unlike the philosophers who use our best science as a window to the world – science as it is practised, as a historical process.

My work falls somewhere in the midst of these three points of departure. My ultimate concern in studying science is with the day-to-day world where SQUIDs can be used to detect stroke victims and where life expectancy is calculated to vary by thirty-five years from one country to another. But the focus of my work is far narrower than that of Dupré: I look at the claims of science, at the possible effects of science as a body of knowledge, in order to see what we can achieve with this knowledge. This puts me much closer to the 'internalist' philosophers in the detail of treatment that I aim for in discussing the image science gives us of the world, but from a different motive. Mine is the motive of the social engineer. Ian Hacking distinguishes two significant aims for science: representing and intervening.[6] Most of my colleagues in philosophy are interested in representing, and not just those specialists whose concerns are to get straight the details of mathematical physics. Consider Bas van Fraassen, who begins with more traditional philosophical worries. Van Fraassen tells us that the foremost question in philosophy of science today is: how can the world be the way science says it is or represents it to be?[7] I am interested in intervening. So I begin from a different question: how can the world be changed by science to make it the way it should be?

The hero behind this book is Otto Neurath, social engineer of the short-lived Bavarian Republic and founding member of the Vienna Circle.[8] Neurath is well known among philosophers for his boat metaphor attacking the

[5] *Cf.* Dupré 1993.
[6] Hacking 1983.
[7] Van Fraassen 1991.
[8] See Cartwright *et al.* 1996.

foundational picture of knowledge. What is less well known is how Neurath went about doing philosophy. He was not concerned with philosophy as a subject in itself. Rather his philosophy was finely tuned to his concerns to change the world. Neurath was the Director of the Agency for Full Social Planning under all three socialist governments in Munich in 1919 and 1920 and he was a central figure in the housing movement and the movement for workers' education in Red Vienna throughout the Vienna Circle period. Neurath advocated what I have learned from my collaborator Thomas Uebel to call *the scientific attitude*. This is the attitude I try to adopt throughout this book. The scientific attitude shares a great deal with conventional empiricism. Most important is the requirement that it is the world around us, the messy, mottled world that we live in and that we wish to improve on, that is the object of our scientific pursuits, the subject of our scientific knowledge, and the tribunal of our scientific judgements. But it departs from a number of empiricisms by rejecting a host of philosophical constructs that are ill-supported by the mottled world in which we live, from Hume's impressions and the inert occurrent properties that replace them in contemporary philosophy to the pseudo-rationalist ideal of universal determinism.

Neurath was the spearhead of the Vienna Circle's Unity of Science Movement. But this does not put us at odds about the patchwork of law. For Neurath worked hard to get us to give up *our belief in the system*. 'The system' for Neurath is the one great scientific theory into which all the intelligible phenomena of nature can be fitted, a unique, complete and deductively closed set of precise statements. Neurath taught:

'The' system is the great scientific lie.[9]

That is what I aim to show in this book. My picture of the relations among the sciences is like Neurath's. Figure 0.1 shows what is often taken to be the standard Vienna Circle doctrine on unity of science: the laws and concepts of each scientific domain are reducible to those of a more fundamental domain, all arranged in a hierarchy, till we reach physics at the pinnacle. Figure 0.2 is Neurath's picture: the sciences are each tied, both in application and confirmation, to the same material world; their language is the shared language of space-time events. But beyond that there is no system, no fixed relations among them. The balloons can be tied together to co-operate in different ways and in different bundles when we need them to solve different problems. Their boundaries are flexible: they can be expanded or contracted; they can even come to cover some of the same territory. But they undoubtedly have boundaries. There is no universal cover of law.

[9] Neurath 1935, p. 116.

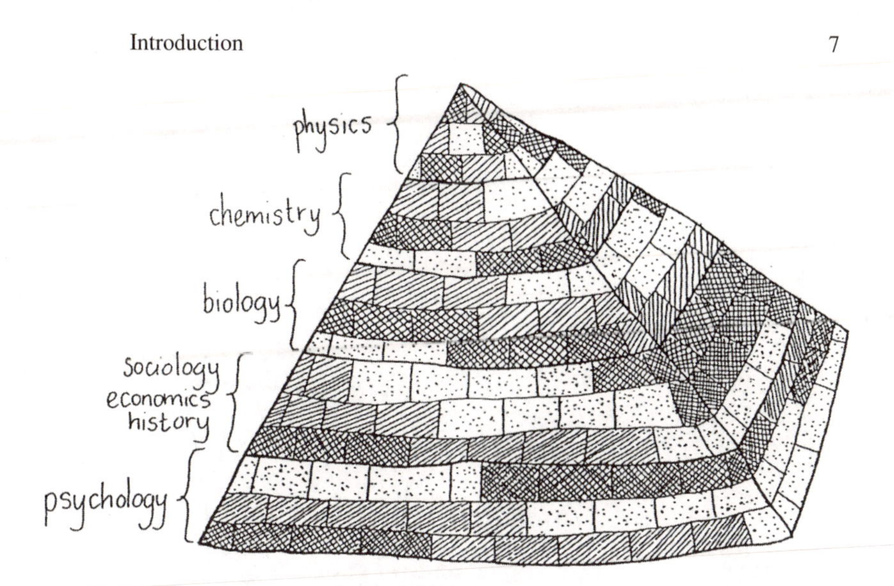

Figure 0.1 Pyramid. Source: Rachel Hacking.

The similarity of my views with Neurath's becomes clear when we turn to questions about the closure of theories. Consider theories in physics like classical Newtonian mechanics, quantum mechanics, quantum field theory, quantum electrodynamics, and Maxwell's theory of electromagnetism, or theories in economics like rational expectations or utility theory, Keynesian macroeconomics or Marxism. In these cases we have been marvellously successful in devising or discovering sets of concepts that have the features traditionally required for scientific knowledge. They are unambiguous: that is, there are clear criteria that determine when they obtain and when not. They are precise: they can not only be ordered as more or less; they can also be given quantitative mathematical representations with nice algebraic and topological features. They are non-modal: they do not refer to facts that involve possibility, impossibility or necessity, nor to ones that involve causality. Finally they have exact relations among themselves, generally expressed in equations or with a probability measure. Something like this last is what we usually mean when we talk of 'closure'.

Exactly what kind of closure do the concepts of our best theories in physics have? The scientific attitude matters here. The kind of closure that is supported by the powerful empirical successes of these theories, I shall argue, is of a narrowly restricted kind: so long as no factors relevant to the effect in question operate except ones that can be appropriately represented by the concepts of the theory, the theory can tell us, to a very high degree of approximation, what the effect will be. But what kinds of factors can be represented

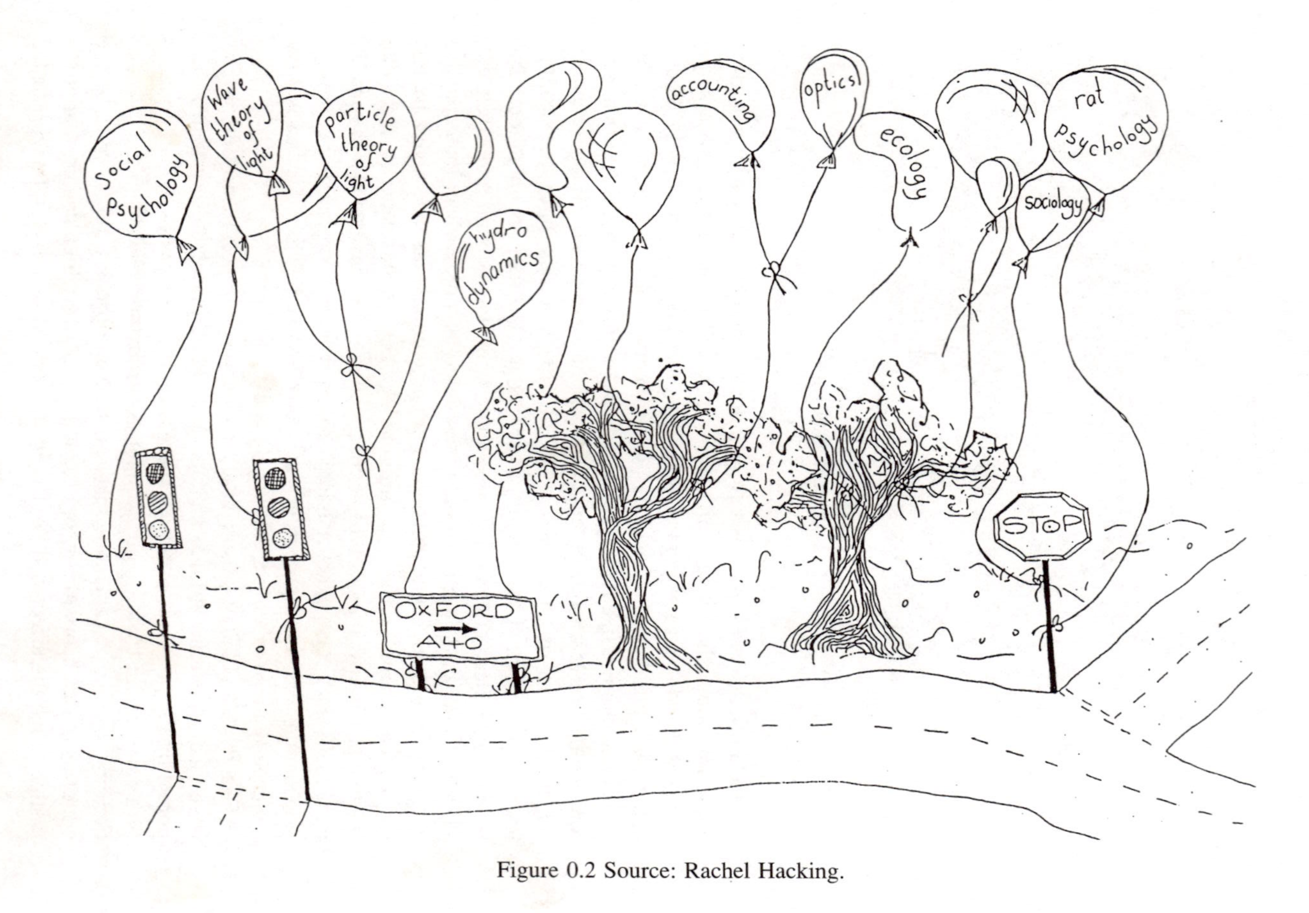

Figure 0.2 Source: Rachel Hacking.

by the concepts of the theory? The interpretative models of the theory provide the answer. And what kinds of interpretative models do we have? In answering this, I urge, we must adopt the scientific attitude: we must look to see what kinds of models our theories have and how they function, particularly how they function when our theories are most successful and we have most reason to believe in them. In this book I look at a number of cases which are exemplary of what I see when I study this question. It is primarily on the basis of studies like these that I conclude that even our best theories are severely limited in their scope. For, to all appearances, not many of the situations that occur naturally in our world fall under the concepts of these theories. That is why physics, though a powerful tool for predicting and changing the world, is a tool of limited utility.

This kind of consideration is characteristic of how I arrive at my image of the dappled world. I take seriously the realists' insistence that where we can use our science to make very precise predictions or to engineer very unnatural outcomes, there must be 'something right' about the claims and practices we employ. I will not in this book go into what the 'something right' could be, about which there is a vast philosophic literature. Rather I want to consider what image of the material world is most consistent with our experiences of it, including our impressive successes at understanding, predicting and manipulating it – but not excluding the limitations within which we find ourselves confined and the repeated failures to get it right that constitute far and away the bulk of normal scientific activity. The logic of the realists' claims is two-edged: if it is the impressive empirical successes of our premier scientific theories that are supposed to argue for their 'truth' (whatever is the favoured interpretation of this claim), then it is the theories as used to generate these empirical successes that we are justified in endorsing.

How do we use theory to understand and manipulate real concrete things – to model particular physical or socio-economic systems? How can we use the knowledge we have encoded in our theories to build a laser or to plan an economy? The core idea of all standard answers is the deductive-nomological account. This is an account that serves the belief in the one great scientific system, a system of a small set of well co-ordinated first principles, admitting a simple and elegant formulation, from which everything that occurs, or everything of a certain type or in a certain category that occurs, can be derived. But treatments of real systems are not deductive; nor are they approximately deductive, nor deductive with correction, nor plausibly approaching closer and closer to deductivity as our theories progress. And this is true even if we tailor our systems as much as possible to fit our theories, which is what we do when we want to get the best predictions possible. That is, it is not true even in the laboratory, as we learn from Peter

Galison, who argues that even the carefully controlled environments of a high-energy physics laboratory do not produce conclusions that can be counted as deductive consequences of physics theories.[10]

What should be put in place of the usual deductive-nomological story then? Surely there is no single account. Occasionally we can produce a treatment that relies on a single theory. Then the application of scientific knowledge can look close to deduction. But my thesis (1) teaches that these deductive accounts will work only in very special circumstances, circumstances that fit the models of the theory in just the right way. Even then, theses (2) and (3) imply that to cash out the *ceteris paribus* conditions of the laws, the deductions will need premises whose language lies outside the given theory; indeed premises that use modal concepts and are thus outside exact science altogether, as exact science is conventionally conceived.

Unfortunately, the special kinds of circumstances that fit the models of a single theory turn out to be hard to find and difficult to construct. More often we must combine both knowledge and technical know-how from a large number of different fields to produce a model that will agree well enough on the matters we are looking to predict, with the method of combination justified at best very locally. And a good deal of our knowledge, as thesis (3) argues, is not of laws but of natures. These tell us what *can* happen, not what will happen, and the step from possibility to actuality is a hypothesis to be tested or a bet to be hedged, not a conclusion to be credited because of its scientific lineage. The point is that the claims to knowledge we can defend by our impressive scientific successes do not argue for a unified world of universal order, but rather for a dappled world of mottled objects.

The conclusions I defend here are a development of those I argued for in *How the Laws of Physics Lie*. They invite the same responses. Realists tend towards universal order, insisting not only that the laws of our best sciences are true or are approaching truth but also that they are 'few in number', 'simple' and 'all-embracing'. *How the Laws of Physics Lie* maintained the opposite: the laws that are the best candidates for being literally true, whether very phenomenological laws or far more abstract ones, are numerous and diverse, complicated and limited in scope. My arguments, then as now, proceeded by looking at our most successful scientific practices. But what can be supported by arguments like these is, as realist-physicist Philip Allport points out, 'a far cry from establishing that such a realist account is impossible'.[11]

My answer to objections like Allport's was already given by David Hume a long time ago. Recall the structure of Hume's *Dialogues Concerning Natural*

[10] See Galison 1987 and 1997.
[11] Allport 1993, p. 254.

Religion. The project is natural religion: to establish the properties that God is supposed to have – omniscience, omnipotence, and benevolence – from the phenomena of the natural world. The stumbling block is evil. Demea and Cleanthes try to explain it away, with well-known arguments. Demea, for example, supposes 'the present evil of phenomena, therefore, are rectified in other regions, and at some future period of existence'.[12] Philo replies:

> I will allow that pain or misery in man is *compatible* with infinite power and goodness in the Deity . . . what are you advanced by all these concessions? A mere possible compatibility is not sufficient. You must *prove* these pure, unmixed and uncontrollable attributes from the present mixed and confused phenomena, and from these alone.[13]

Philo expands his argument:

> [I]f a very limited intelligence whom we shall suppose utterly unacquainted with the universe were assured that it were the production of a very good, wise, and powerful being, however finite, he would, from his conjecture, form beforehand a very different notion of it from what we find it to be by experience; . . . supposing now that this person were brought into the world, still assured that it was the workmanship of such a divine and benevolent being, he might, perhaps, be surprised at the disappointment, but would never retract his former belief if founded on any very solid argument . . . But suppose, which is the real case with regard to man, that this creature is not antecedently convinced of a supreme intelligence, benevolent and powerful, but is left to gather such a belief from the appearance of things; this entirely alters the case, nor will he ever find any reason for such a conclusion.[14]

Philo is the self-avowed mystic. He believes in the wondrous characteristics of the deity but he takes it to be impossible to defend this belief by reason and evidence. Philo's is not natural religion, but revealed.

Hume's project was natural religion. My project is natural science. So too, I take it, is the project of Allport and others who join him in the belief in universal order. I agree that my illustrations of the piecemeal and bitty fashion in which our most successful sciences operate are 'a far cry' from showing that the system must be a great scientific lie. But I think we must approach natural science with at least as much of the scientific attitude as natural religion demands. Complication and limitation in the truest laws we have available are compatible with simplicity and universality in the unknown ultimate laws. But what is advanced by this concession? Just as we know a set of standard moves to handle the problem of evil, so too are we well rehearsed in the problem of unruliness in nature – and a good number of the replies have the same form in both discourses: the problem is not in nature but,

[12] Hume 1779 [1980], part X.
[13] Hume 1779 [1980], part X.
[14] Hume 1779 [1980], part XI.

rather, an artefact of our post-lapsarian frailties. I follow Philo in my reply: guarantee nothing *a priori*, and gather our beliefs about laws, if we must have them at all, from the appearance of things.

The particular theory in dispute between Allport and me was the Standard Model for fundamental particles. Let us allow, for the sake of argument, that the Standard Model can provide a good account of our experimental evidence, that is, that our experimental models for the fundamental fermions and bosons which concern Allport fit exactly into the Standard Model without any of the kinds of fudging or distortion that I worried about in *How the Laws of Physics Lie*. Since I have no quarrel with induction as a form of inference, I would be willing to take this as good evidence that the Standard Model is indeed true[15] of fundamental bosons and fermions – in situations relevantly similar to those of our experiments. Does that make it all-embracing? That depends on how much of the world is 'relevantly similar'. And that is a matter for hard scientific investigation, not *a priori* metaphysics. That is the reason I am so concerned with the successes and failures of basic science in treating large varieties of situations differing as much as possible from our experimental arrangements.

As I have indicated, my investigations into how basic science works when it makes essential contributions to predicting and rebuilding the world suggest that even our best theories are severely limited in their scope: they apply only in situations that resemble their models, and in just the right way, where what constitutes a model is delineated by the theory itself. Classical mechanics, for instance, can deal with small compact masses, rigid rods and point charges; but it is not very successful with things that are floppy, elastic or fluid. Still, it should be clear from my discussion of natural religion, natural science and the scientific attitude that any conclusion we draw from this about the overall structure and extent of the laws of nature must be a guarded one. The dappled world that I describe is best supported by the evidence, but it is clearly not compelled by it.

Why then choose at all? Or, why not choose the risky option, the world of unity, simplicity and universality? If nothing further were at stake, I should not be particularly concerned about whether we believe in a ruly world or in an unruly one, for, not prizing the purity of our affirmations, I am not afraid that we might hold false beliefs. The problem is that our beliefs about the structure of the world go hand-in-hand with the methodologies we adopt to study it. The worry is not so much that we will adopt wrong images with which to represent the world, but rather that we will choose wrong tools with which to change it. We yearn for a better, cleaner, more orderly world than the one that, to all appearances, we inhabit. But it will not do to base our

[15] In the same sense in which we judge other general claims true or false, whatever sense that is.

Figure 0.3 Source: Rachel Hacking.

methods on our wishes. We had better choose the most probable option and wherever possible hedge our bets.

A series of pictures drawn by Rachel Hacking illustrates my overall thesis. The figure in the pictures represents the philosopher or scientist who is yearning for unity and simplicity. In figure 0.3 he stands in the unruly and untidy world around him with a wistful dream of a better world elsewhere. Notice that our hero is about to make a leap. Figure 0.4 shows the ideal world he longs for. Figure 0.5 is the messy and ill-tended world that he really inhabits. Figure 0.6 is the key to my worries. It shows the disaster that our hero creates when he tries to remake the world he lives in following the model of his ideal image. Figure 0.7 is what I urge: it shows what can be accomplished if improvements are guided by what is possible not what is longed for.

For a concrete example of how belief in the unity of the world and the completeness of theory can lead to poor methodology we can turn to a discussion by Daniel Hausman of why evidence matters so little to economic theory.[16] Hausman notices that economists studying equilibrium theory make serious efforts to improve techniques for gathering and analysing market data, but they do not generally seek or take seriously other forms of data, particu-

[16] Hausman 1997.

Figure 0.4 Source: Rachel Hacking.

Figure 0.5 Source: Rachel Hacking.

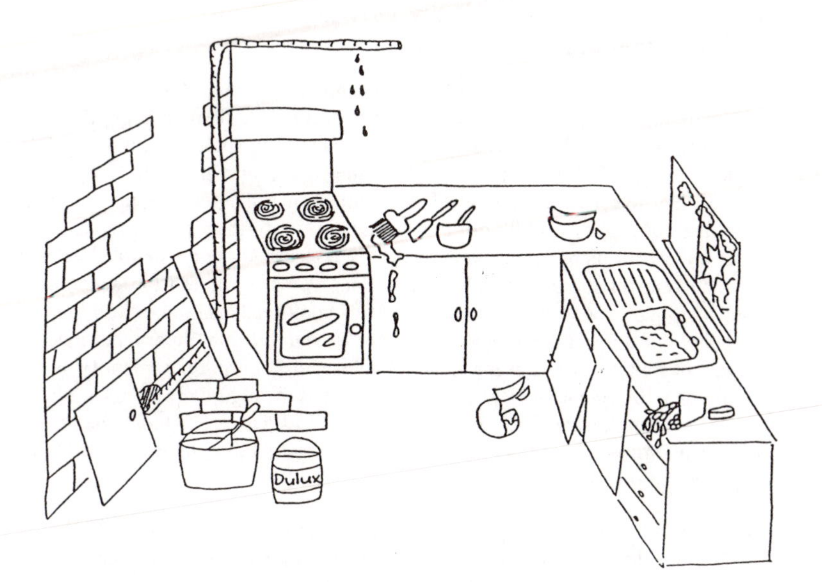

Figure 0.6 Source: Rachel Hacking.

Figure 0.7 Source: Rachel Hacking.

larly data that might be supplied from psychology experiments. Nor do they use the results of 'high quality tests' in their own field to improve their theories. Here is Hausman's account of why:

[M]any economists believe that equilibrium theory provides, at a certain level of resolution, a complete theory of the whole economic domain. Equilibrium theorists rarely state this thesis explicitly, for on the face of it, such a claim is highly implausible. But equilibrium economists show their commitment to this implausible view when they reject as *ad hoc* any behavioural generalizations about individual agents that are not part of equilibrium theory and are not further specifications of the causal factors with which equilibrium theory is concerned. For example, as part of his explanation for the existence of depressions, Keynes offered the generalization that the marginal propensity to consume out of additional income is less than one. This psychological generalization is consistent with equilibrium theory, but independent: it does not follow from equilibrium theory. Instead, it purports to identify an *additional* important causal factor. For precisely this reason, regardless of its truth or falsity, many equilibrium theorists have rejected it as *ad hoc* . . . If those theorists believed that economists need to uncover additional causal factors, then they would not reject attempts to do so.[17]

Economics is not special in this respect. The pernicious effects of the belief in the single universal rule of law and the single scientific system are spread across the sciences. Indeed, a good many physicists right now are in revolt. Superstring theory is the new candidate for a theory of everything.[18] P. W. Anderson is one of its outspoken opponents. The theory consumes resources and efforts that could go into the hundreds of other enterprises in physics that ask different kinds of questions and solve different kinds of problems. The idea that solutions can trickle down from superstring theory is not even superficially plausible. Anderson urges,

The ability to reduce everything to simple fundamental laws does not imply the ability to start from those laws and reconstruct the universe. In fact, the more the elementary particle physicists tell us about the nature of the fundamental laws, the less relevance they seem to have to the very real problems of the rest of science, much less to those of society . . . [T]he behavior of large and complex aggregates of elementary particles, it turns out, is not to be understood in terms of a simple extrapolation of the properties of a few particles. Instead, at each level of complexity entirely new properties appear, and the understanding of the new behaviors requires research which I think is as fundamental in its nature as any other.[19]

The damage from the move for unity is not done when there is good empirical evidence for a specific kind of unification. Nor are attempts at take-overs out

[17] *Ibid*., p. 406.

[18] For a full discussion of unificationalist programmes in theoretical physics and what they amount to I recommend Margaret Morrison's study *Unifying Theories*, Morrison forthcoming.

[19] Anderson 1972, p. 393; for a discussion of Anderson's recent attacks, see Cat 1998.

of place when we have good reason to think the methods or ideas we have developed to solve one kind of problem can be employed to solve a different kind, when we can argue the case and make a reasonable bet, where the costs have been thought through and our assessments of the chances for success can warrant this way of proceeding over others. If it is not even objectionable to invest in a hunt for unity out of sheer passion for order and rationality, nor to try our favoured theory just because we have no better ideas or because that is what we are good at. It is not objectionable so long as we are clear about what we arc doing and what it is going to cost us, and we are able to pay the price.

But we are often not so clear. The yearning for 'the system' is a powerful one; the faith that our world must be rational, well ordered through and through, plays a role where only evidence should matter. Our decisions are affected. After the evidence is in, theories that purport to be fundamental – to be able in principle to explain everything of a certain kind – often gain additional credibility just for that reason itself. They get an extra dollop of support beyond anything they have earned by their empirical success or the empirically warranted promise of their research programme for solving the problem at hand. I have mentioned already the take-over attempts of strings and symmetries and dualities, of fundamental particles in physics, of quantum mechanics over classical, and equilibrium theory, rational expectations and game theory in political economy. In medicine one of the primary take-over theories is genetics. Let us consider it.

Genetics has proved very useful across a variety of problems. Genetic information about the heritability of the production of Down's syndrome babies has led to much better targeting of *amniocentesis* among young mothers. Our knowledge that glycogen storage diseases are due to single point mutations allows us to administer the somewhat dangerous tests for these diseases only to babies with a family history. Now that we understand that the serious mental handicaps of phenylketonuria (PKU) are due to a single point mutation that leads too much accumulation of one amino acid and not enough of another with a resulting failure of neurological development, we can adjust the diet of the children affected till the relevant period of development is over. Some even hope that we will learn how to use viruses as vectors to resupply the genetic material.

For many diseases, however, other approaches may be just as, or even more, fruitful. For example, in breast cancer there is very substantial evidence that endogenous oestrogen levels are the major determining factor in the occurrence of the disease in the vast majority of cases. It is well known that endogenous oestrogen levels are affected by lifestyle, but little emphasis is put on this aspect of prevention: on finding out how diet, physical activity (exercise and work) and other modifiable factors may lower endogenous oestrogen levels, or on finding 'synthetic' approaches that could achieve the same end. A rational approach to the problem of breast cancer prevention would put as much emphasis here as on the genetic aspects.[20]

Is the level of effort and funding that goes into the gene programme versus the others warranted by the promise of these programmes for understanding and controlling breast cancer or does the gene programme get a substantial edge because it is the gene programme; because it is our best shot right now at a theory of everything? I care about our ill supported beliefs that nature is governed by some universal theories because I am afraid that women are dying of breast cancer when they need not do so because other programmes with good empirical support for their proposals are ignored or underfunded. On the side of political economy, Chicago economists carry the conclusions of the putatively all-encompassing theory of rational expectations far, far away from the models where they can derive them to admonish the government against acting to improve social welfare; and the World Bank and the International Monetary Fund use statistical models of the kind I criticise in chapter 7 to discourage Third World countries from direct welfare expenditure. That is what is wrong with faith in the great scientific system.

My own research right now is not primarily concerned with economics or with physics or with any other single discipline from within. It is concerned rather with how to get the most out of our scientific knowledge as a whole. How do we best put together different levels and different kinds of knowledge from different fields to solve real world problems, the bulk of which do not fall in any one domain of any one theory? Within each of the disciplines separately, both pure and applied, we find well developed, detailed methodologies both for judging claims to knowledge and for putting them to use. But we have no articulated methodologies for interdisciplinary work, not even anything so vague and general as the filtered-down versions of good scientific method that we are taught at school. To me this is the great challenge that now faces philosophy of science: to develop methodologies, not for life in the laboratory where conditions can be set as one likes, but methodologies for life in the messy world that we inevitably inhabit.

I am sometimes told that my project is unexciting and uninspiring, for both philosophers and scientists alike. Consider again Philip Allport, who contrasts my view with Karl Popper's. Allport argues that it is Popper's 'lofty vision' as opposed to my 'more practical' and 'bottom up' perspective that will 'communicate the excitement of pure research and the beauty of the theoretical vision that motivates it'.[21] That, I think, depends entirely on where one finds beauty. Popper and Allport, it seems, find beauty in 'the mastery of the

[20] Thanks to Malcolm Pike for his help with this example.
[21] Allport 1991, p. 56.

whole world of experience, by subsuming it ultimately under one unified theoretical structure'.[22] I follow Gerard Manley Hopkins in feeling the beauty of Pied Beauty:

> GLORY be to God for dappled things–
> For skies of couple-colour as a brindled cow;
> For rose-moles all in stipple upon trout that swim;
> Fresh-firecoal chestnut-falls; finches' wings;
> Landscape plotted and pieced – fold, fallow, and plough;
> And áll trádes, their gear and tackle and trim.
> All things counter, original, spare, strange;
> Whatever is fickle, freckled (who knows how?)
> With swift, slow; sweet, sour; adazzle, dim;
> He fathers-forth whose beauty is past change:
> Praise him.

[22] *Ibid.*, Allport is here quoting Gerald Holton.

Part I

Where do laws of nature come from?

1 Fundamentalism versus the patchwork of laws

1 For realism

A number of years ago I wrote *How The Laws of Physics Lie*. That book was generally perceived to be an attack on realism. Nowadays I think that I was deluded about the enemy: it is not *realism* but *fundamentalism* that we need to combat.

My advocacy of realism – local realism about a variety of different kinds of knowledge in a variety of different domains across a range of highly differentiated situations – is Kantian in structure. Kant frequently used a puzzling argument form to establish quite abstruse philosophical positions (Φ): We have X – perceptual knowledge, freedom of the will, whatever. But without Φ (the transcendental unity of the apperception, or the kingdom of ends) X would be impossible, or inconceivable. Hence Φ. The objectivity of local knowledge is my Φ; X is the possibility of planning, prediction, manipulation, control and policy setting. Unless our claims about the expected consequences of our actions are reliable, our plans are for nought. Hence knowledge is possible.

What might be found puzzling about the Kantian argument form are the Xs from which it starts. These are generally facts which appear in the clean and orderly world of pure reason as refugees with neither proper papers nor proper introductions, of suspect worth and suspicious origin. The facts which I take to ground objectivity are similarly alien in the clear, well-lighted streets of reason, where properties have exact boundaries, rules are unambiguous, and behaviour is precisely ordained. I know that I can get an oak tree from an acorn, but not from a pine cone; that nurturing will make my child more secure; that feeding the hungry and housing the homeless will make for less misery; and that giving more smear tests will lessen the incidence of cervical cancer. Getting closer to physics, which is ultimately my topic here, I also know that I can drop a pound coin from the upstairs window into the hands of my daughter below, but probably not a paper tissue; that I can head north by following my compass needle (so long as I am on foot and not in my car), that . . .

I know these facts even though they are vague and imprecise, and I have no reason to assume that that can be improved on. Nor, in many cases, am I sure of the strength or frequency of the link between cause and effect, nor of the range of its reliability. And I certainly do not know in any of the cases which plans or policies would constitute an optimal strategy. But I want to insist that these items are items of knowledge. They are, of course, like all genuine items of knowledge (as opposed to fictional items like sense data or the synthetic *a priori*) defeasible and open to revision in the light of further evidence and argument. But if I do not know these things, what do I know and how can I come to know anything?

Besides this odd assortment of inexact facts, we also have a great deal of very precise and exact knowledge, chiefly supplied by the natural sciences. I am not thinking here of abstract laws, which as an empiricist I take to be of considerable remove from the world they are supposed to apply to, but rather of the precise behaviour of specific kinds of concrete systems, knowledge of, say, what happens when neutral *K*-mesons decay, which allows us to establish CP violation, or of the behaviour of SQUIDs (Superconducting Quantum Interference Devices) in a shielded fluctuating magnetic field, which allows us to detect the victims of strokes. This knowledge is generally regimented within a highly articulated, highly abstract theoretical scheme.

One cannot do positive science without the use of induction, and where those concrete phenomena can be legitimately derived from the abstract schemes, they serve as a kind of inductive base for these schemes. *How The Laws of Physics Lie* challenged the soundness of these derivations and hence of the empirical support for the abstract laws. I still maintain that these derivations are shaky, but that is not the point I want to make in this chapter. So let us for the sake of argument assume the contrary: the derivations are deductively correct and they use only true premises. Then, granting the validity of the appropriate inductions,[1] we have reason to be realists about the laws in question. But that does not give us reason to be fundamentalists. To grant that a law is true – even a law of 'basic' physics or a law about the so-called 'fundamental particles' – is far from admitting that it is universal – that it holds everywhere and governs in all domains.

2 Against fundamentalism

Return to my rough division of the concrete facts we know into two categories: (1) those that are legitimately regimented into theoretical schemes, these generally, though not always, being about behaviour in highly structured, manufactured environments like a spark chamber; (2) those that are not.

[1] These will depend on the circumstances and our general understanding of the similarities and structures that obtain in those circumstances.

There is a tendency to think that all facts must belong to one grand scheme, and moreover that this is a scheme in which the facts in the first category have a special and privileged status. They are exemplary of the way nature is supposed to work. The others must be made to conform to them. This is the kind of fundamentalist doctrine that I think we must resist. Biologists are clearly already doing so on behalf of their own special items of knowledge. Reductionism has long been out of fashion in biology and now emergentism is again a real possibility. But the long-debated relations between biology and physics are not good paradigms for the kind of anti-fundamentalism I urge. Biologists used to talk about how new laws emerge with the appearance of 'life'; nowadays they talk, not about life, but about levels of complexity and organisation. Still in both cases the relation in question is that between larger, richly endowed, complex systems, on the one hand, and fundamental laws of physics on the other: it is the possibility of 'downwards' reduction that is at stake.

I want to go beyond this. Not only do I want to challenge the possibility of downwards reduction but also the possibility of 'cross-wise reduction'. Do the laws of physics that are true of systems (literally true, we may imagine for the sake of argument) in the highly contrived environments of a laboratory or inside the housing of a modern technological device, do these laws carry across to systems, even systems of very much the same kind, in different and less regulated settings? Can our refugee facts always, with sufficient effort and attention, be remoulded into proper members of the physics community, behaving tidily in accord with the fundamental code? Or must – and should – they be admitted into the body of knowledge on their own merit?

In moving from the physics experiment to the facts of more everyday experience, we are not only changing from controlled to uncontrolled environments, but often from micro to macro as well. In order to keep separate the issues which arise from these two different shifts, I am going to choose for illustration a case from classical mechanics, and will try to keep the scale constant. Classical electricity and magnetism would serve as well. Moreover, in order to make my claims as clear as possible, I shall consider the simplest and most well-known example, that of Newton's second law and its application to falling bodies, $\mathbf{F} = m\mathbf{a}$. Most of us, brought up within the fundamentalist canon, read this with a universal quantifier in front: for any body in any situation, the acceleration it undergoes will be equal to the force exerted on it in that situation divided by its inertial mass. I want instead to read it, as indeed I believe we should read *all* nomologicals, as a *ceteris paribus* law. In later chapters I shall have a lot to say about the form of this *ceteris paribus* condition. Indeed, I shall argue that laws like this are very special indeed and they obtain only in special circumstances: they obtain just when a nomological machine is at work. But for the moment, to get started, let us concentrate on the more usual observation that, for the most part, the relations laid out in our laws hold only if nothing untoward intervenes to prevent them. In our example, then, we may write: for any body in any situation, *if nothing interferes*, its acceleration will equal the force exerted on it divided by its mass. But what can interfere with a force in the production of motion other than another force? Surely there is no problem. The acceleration will always be equal to the *total* force divided by the mass. That is just what I question.

Think again about how we construct a theoretical treatment of a real situation. Before we can apply the abstract concepts of basic theory – assign a quantum field, a tensor, a Hamiltonian, or in the case of our discussion, write down a force function – we must first produce a model of the situation in terms the theory can handle. From that point the theory itself provides 'language-entry rules' for introducing the terms of its own abstract vocabulary, and thereby for bringing its laws into play. *How The Laws of Physics Lie* illustrated this for the case of the Hamiltonian – which is roughly the quantum analogue of the classical force function. Part of learning quantum mechanics is learning how to write the Hamiltonian for canonical models, for example, for systems in free motion, for a square well potential, for a linear harmonic oscillator, and so forth. Ronald Giere has made the same point for classical mechanics.[2] In the next chapter I will explain that there is a very special kind of abstraction that is at stake here. For now let us consider what follows from the fact that concepts like *force* or the quantum Hamiltonian attach to the world through a set of specific models. (I develop a quantum example in detail in chapter 9.)

The basic strategy for treating a real situation is to piece together a model from these fixed components. Then we determine the prescribed composite Hamiltonian from the Hamiltonians for the parts. Questions of realism arise when the model is compared with the situation it is supposed to represent. *How the Laws of Physics Lie* argued that even in the best cases, the fit between the two is not very good. I concentrated there on the best cases because I was trying to answer the question 'Do the explanatory successes of modern theories argue for their truth?' Here I want to focus on the multitude of 'bad' cases, where the models, if available at all, provide a very poor image of the situation. These are not cases that disconfirm the theory. You cannot show that the predictions of a theory for a given situation are false until you have managed to describe the situation in the language of the theory. When the models are too bad a fit, the theory is not disconfirmed; it is just inapplicable.

Now consider a falling object. Not Galileo's from the leaning tower, nor the pound coin I earlier described dropping from the upstairs window, but

[2] Giere 1988.

rather something more vulnerable to non-gravitational influence. Otto Neurath has a nice example. My doctrine about the case is much like his.

> In some cases a physicist is a worse prophet than a [behaviourist psychologist], as when he is supposed to specify where in Saint Stephen's Square a thousand dollar bill swept away by the wind will land, whereas a [behaviourist] can specify the result of a conditioning experiment rather accurately.[3]

Mechanics provides no model for this situation. We have only a partial model, which describes the thousand dollar bill as an unsupported object in the vicinity of the earth, and thereby introduces the force exerted on it due to gravity. Is that the total force? The fundamentalist will say *no*: there is in principle (in God's completed theory?) a model in mechanics for the action of the wind, albeit probably a very complicated one that we may never succeed in constructing. This belief is essential for the fundamentalist. If there is no model for the thousand dollar bill in mechanics, then what happens to the note is not determined by its laws. Some falling objects, indeed a very great number, will be outside the domain of mechanics, or only partially affected by it. But what justifies this fundamentalist belief? The successes of mechanics in situations that it can model accurately do not support it, no matter how precise or surprising they are. They show only that the theory is true in its domain, not that its domain is universal. The alternative to fundamentalism that I want to propose supposes just that: mechanics is true, literally true we may grant, for all those motions all of whose causes can be adequately represented by the familiar models which get assigned force functions in mechanics. For these motions, mechanics is a powerful and precise tool for prediction. But for other motions, it is a tool of limited serviceability.

Let us set our problem of the thousand dollar bill in Saint Stephen's Square to an expert in fluid dynamics. The expert should immediately complain that the problem is ill defined. What exactly is the bill like: is it folded or flat? straight down the middle? or . . .? is it crisp or crumpled? how long versus wide? and so forth and so forth and so forth. I do not doubt that *when* the right questions can be asked and the right answers can be supplied, fluid dynamics can provide a practicable model. But I do doubt that for every real case, or even for the majority, fluid dynamics has enough of the 'right questions'. It does not have enough of the right concepts to allow it to model the full set of causes, or even all the dominant ones. I am equally sceptical that the models which work will do so by legitimately bringing Newton's laws (or Lagrange's for that matter) into play.[4]

How then do airplanes stay aloft? Two observations are important. First,

[3] Neurath 1933, p. 13.

[4] And the problem is certainly not that a quantum or relativistic or microscopic treatment is needed instead.

we do not need to maintain that no laws obtain where mechanics runs out. Fluid dynamics may have loose overlaps and intertwinings with mechanics. But it is in no way a subdiscipline of basic classical mechanics; it is a discipline on its own. Its laws can direct the thousand dollar bill as well as can those of Newton or Lagrange. Second, the thousand dollar bill comes as it comes, and we have to hunt a model for it. Just the reverse is true of the plane. We build it to fit the models we know work. Indeed, that is how we manage to get so much into the domain of the laws we know.

Many will continue to feel that the wind and other exogenous factors must produce a force. The wind after all is composed of millions of little particles which must exert all the usual forces on the bill, both across distances and via collisions. That view begs the question. When we have a good-fitting molecular model for the wind, and we have in our theory (either by composition from old principles or by the admission of new principles) systematic rules that assign force functions to the models, and the force functions assigned predict exactly the right motions, then we will have good scientific reason to maintain that the wind operates via a force. Otherwise the assumption is another expression of fundamentalist faith.

3 *Ceteris Paribus* laws versus ascriptions of natures

If the laws of mechanics are not universal, but nevertheless true, there are at least two options for them. They could be pure *ceteris paribus* laws: laws that hold only in circumscribed conditions or so long as no factors relevant to the effect besides those specified occur. And that's it. Nothing follows about what happens in different settings or in cases where other causes occur that cannot be brought under the concepts of the theory in question. Presumably this option is too weak for our example of Newtonian mechanics. When a force is exerted on an object, the force will be relevant to the motion of the object even if other causes for its motion not renderable as forces are at work as well, and the exact relevance of the force will be given by the formula $\mathbf{F} = m\mathbf{a}$: the (total) force will contribute a component to the acceleration determined by this formula.

For cases like this, the older language of *natures* is appropriate. It is in the nature of a force to produce an acceleration of the requisite size. That means that *ceteris paribus*, it will produce that acceleration. But even when other causes are at work, it will 'try' to do so. The idea is familiar in the case of forces. What happens when several forces all try at once to produce their own characteristic acceleration? What results is an acceleration equal to the vector sum of the functions that represent each force separately, divided by the inertial mass of the accelerating body. In general what counts as 'trying' will differ from one kind of cause to another. To ascribe a behaviour to the nature of a feature is to claim that that behaviour is exportable beyond the strict confines of the *ceteris paribus* conditions, although usually only as a 'tendency' or a 'trying'. The extent and range of the exportability will vary. Some natures are highly stable; others are very restricted in their range. The point here is that we must not confuse a wide-ranging nature with the universal applicability of the related law. To admit that forces tend to cause the prescribed acceleration (and indeed do so in felicitous conditions) is a long way from admitting that $\mathbf{F} = m\mathbf{a}$, read as a claim of regular association, is universally true.[5] In the next sections I will describe two different metaphysical pictures in which fundamentalism about the experimentally derived laws of basic physics would be a mistake. The first is wholism; the second, pluralism. It seems to me that wholism is far more likely to give rise only to *ceteris paribus* laws, whereas natures are more congenial to pluralism.

4 Wholism

We look at little bits of nature, and we look under a very limited range of circumstances. This is especially true of the exact sciences. We can get very precise outcomes, but to do so we need very tight control over our inputs. Most often we do not control them directly, one by one, but rather we use some general but effective form of shielding. I know one experiment that aims for direct control – the Stanford Gravity Probe. Still, in the end, they will roll the space ship to average out causes they have not been able to command. Sometimes we take physics outside the laboratory. Then shielding becomes even more important. SQUIDs can make very fine measurements of magnetic fluctuations that help in the detection of stroke victims. But for administering the tests, the hospital must have a Hertz box – a small totally metal room to block out magnetism from the environment. Or, for a more homely example, we all know that batteries are not likely to work if their protective casing has been pierced.

We tend to think that shielding does not matter to the laws we use. The same laws apply both inside and outside the shields; the difference is that inside the shield we know how to calculate what the laws will produce, but outside, it is too complicated. Wholists are wary of these claims. If the events we study are locked together and changes depend on the total structure rather than the arrangement of the pieces, we are likely to be very mistaken by looking at small chunks of special cases.

Goodman's new riddle of induction provides a familiar illustration.[6] Sup-

[5] I have written more about the two levels of generalisation, laws and ascriptions of natures, in Cartwright 1989.

[6] Goodman 1983, ch. 3.

pose that in truth all emeralds are grue (where *grue* = green and examined before the year 2000 or blue and either unexamined or examined after the year 2000). We, however, operate with 'All emeralds are green.' Our law is not an approximation before 2000 of the true law, though we can see from the definitions of the concepts involved why it will be every bit as successful as the truth, and also why it will be hard, right now, to vary the circumstances in just the right way to reverse our erroneous induction.

A second famous example of erroneous induction arising from too narrow a domain of experience is Bertrand Russell's ignorant chicken. The chicken waits eagerly for the arrival of the farmer, who comes to feed her first thing every morning – until the unexpected day that he comes to chop off her head. The chicken has too limited a view of the total situation in which she is embedded. The entire structure is different from what she suspects, and designed to serve very different purposes from her own. Yet, for most of her life, the induction that she naturally makes provides a precisely accurate prediction.

For a more scientific example consider the revolution in communications technology due to fibre optics. Low-loss optical fibres can carry information at rates of many gigabits per second over spans of tens of kilometres. But the development of fibre bundles which lose only a few decibels per kilometre is not all there is to the story. Pulse broadening effects intrinsic to the fibres can be truly devastating. If the pulses broaden as they travel down the fibre, they will eventually smear into each other and destroy the information. That means that the pulses cannot be sent too close together, and the transmission rate may drop to tens or at most hundreds of megabits per second.

We know that is not what happens. The technology has been successful. That is because the right kind of optical fibre in the right circumstance can transmit solitons – solitary waves that keep their shape across vast distances. I will explain why. The light intensity of the incoming pulse causes a shift in the index of refraction of the optical fibre, producing a slight non-linearity in the index. The non-linearity leads to what is called a 'chirp' in the pulse. Frequencies in the leading half of the pulse are lowered while those in the trailing half are raised. The effects of the chirp combine with those of dispersion to produce the soliton. Thus stable pulse shapes are not at all a general phenomenon of low-loss optical fibres. They are instead a consequence of two different, oppositely directed processes that cancel each other out. The pulse widening due to the dispersion is cancelled by the pulse narrowing due to the non-linearity in the index of refraction. We can indeed produce perfectly stable pulses. But to do so we must use fibres of just the right design, and matched precisely with the power and input frequency of the laser which generates the input pulses. By chance that was not hard to do. When the ideas were first tested in 1980 the glass fibres and lasers readily available were easily suited to each other. Given that very special match, fibre optics was off to an impressive start.

Solitons are indeed a stable phenomenon. They are a feature of nature, but of nature under very special circumstances. Clearly it would be a mistake to suppose that they were a general characteristic of low-loss optical fibres. The question is, how many of the scientific phenomena we prize are like solitons, local to the environments we encounter, or – more importantly – to the environments we construct. If nature is more wholistic than we are accustomed to think, the fundamentalists' hopes to export the laws of the laboratory to the far reaches of the world will be dashed.

It is clear that I am not very sanguine about the fundamentalist faith. But that is not really out of the kind of wholist intuitions I have been sketching. After all, the story I just told accounts for the powerful successes of the 'false' local theory – the theory that solitons are characteristic of low-loss fibres – by embedding it in a far more general theory about the interaction of light and matter. Metaphysically, the fundamentalist is borne out. It may be the case that the successful theories we have are limited in their domain, but their successes are to be explained by reference to a truly universal authority. I do not see why we need to explain their successes. I am prepared to believe in more general theories when we have direct empirical evidence for them but not merely because they are the 'best explanation' for something which may well have no explanation. 'The theory is successful in its domain': the need for explanation is the same whether the domain is small, or large, or very small, or very large. Theories are successful where they are successful, and that's that. If we insist on turning this into a metaphysical doctrine, I suppose it will look like metaphysical pluralism, to which I now turn.

5 The patchwork of laws

Metaphysical nomological pluralism is the doctrine that nature is governed in different domains by different systems of laws not necessarily related to each other in any systematic or uniform way; by a patchwork of laws. Nomological pluralism opposes any kind of fundamentalism. I am here concerned especially with the attempts of physics to gather all phenomena into its own abstract theories. In *How the Laws of Physics Lie* I argued that most situations are brought under a law of physics only by distortion, whereas they can often be described fairly correctly by concepts from more phenomenological laws. The picture suggested was of a lot of different situations in a continuum from ones that fit not perfectly but not badly to those that fit very badly indeed. I did suggest that at one end fundamental physics might run out entirely ('What is the value of the electric field vector in the region just at the tip of my pencil?'), whereas in transistors it works quite well. But that was not the principal focus. Now I want to draw the divides sharply. Some features of systems typically studied by physics may get into situations where their behaviour is not governed by the laws of physics at all. But that does not mean that they have no guide for their behaviour or only low-level phenomenological laws. They may fall under a quite different organised set of highly abstract principles. (But then again they may not.)

There are two immediate difficulties that metaphysical pluralism encounters. The first is one we create ourselves, by imagining that it must be joined with views that are vestiges of metaphysical monism. The second is, I believe, a genuine problem that nature must solve (or must have solved).

First, we are inclined to ask, 'How can there be motions not governed by Newton's laws?' The answer: there are causes of motion not included in Newton's theory. Many find this impossible because, although they have forsaken reductionism, they cling to a near-cousin: *supervenience*. Suppose we give a complete 'physics' description of the falling object and its surrounds. Mustn't that fix all the other features of the situation? Why? This is certainly not true at the level of discussion at which we stand now: the wind is cold and gusty; the bill is green and white and crumpled. These properties are independent of the mass of the bill, the mass of the earth, the distance between them.

I suppose, though, I have the supervenience story wrong. It is the microscopic properties of physics that matter; the rest of reality supervenes on them. Why should I believe that? Supervenience is touted as step forward over reductionism. Crudely, I take it, the advantage is supposed to be that we can substitute a weaker kind of reduction, 'token-token reduction', for the more traditional 'type-type reductions' which were proving hard to carry out. But the traditional view had arguments in its favour. Science does sketch a variety of fairly systematic connections between micro-structures and macro-properties. Often the sketch is rough, sometimes it is precise, usually its reliability is confined to very special circumstances. Nevertheless there are striking cases. But these cases support type-type reductionism; they are irrelevant for supervenience. Type-type reductionism has well-known problems: the connections we discover often turn out to look more like causal connections than like reductions; they are limited in their domain; they are rough rather than exact; and often we cannot even find good starting proposals where we had hoped to produce nice reductions. These problems suggest modifying the doctrine in a number of specific ways, or perhaps giving it up altogether. But they certainly do not leave us with token-token reductionism as a fallback position. After all, on the story I have just told, it was the appearance of some degree of systematic connection that argued in the first place for the claim that micro-structures fixed macro-properties. But it is just this systematicity that is missing in token-token reductionism.

The view that there are macro-properties that do not supervene on micro-features studied by physics is sometimes labelled *emergentism*. The suggestion is that, where there is no supervenience, macro-properties must miraculously come out of nowhere. But why? There is nothing of the newly landed about these properties. They have been here in the world all along, standing right beside the properties of microphysics. Perhaps we are misled by the feeling that the set of properties studied by physics is complete. Indeed, I think that there is a real sense in which this claim is true, but that sense does not support the charge of emergentism. Consider how the domain of properties for physics gets set. Here is a caricature: we begin with an interest in some specific phenomena, say motions – deflections, trajectories, orbits. Then we look for the smallest set of properties that is, *ceteris paribus* closed (or, closed enough) under prediction. That is, we expand our domain until we get a set of factors that are *ceteris paribus* sufficient to fix our starting factors. (That is, they are sufficient so long as nothing outside the set occurs that strongly affects the targeted outcome.) To succeed does not show that we have got all the properties there are. This is a fact we need to keep in mind quite independently of the chief claim of this chapter, that the predictive closure itself only obtains in highly restricted circumstances. The immediate point is that predictive closure among a set of properties does not imply descriptive completeness.

The second problem that metaphysical pluralism faces is that of consistency. We do not want colour patches to appear in regions from which the laws of physics have carried away all matter and energy. Here are two stories I used to tell when I taught about the Mechanical Philosophy of the seventeenth century. Both are about how to write the Book of Nature to guarantee consistency. In the first story, God is very interested in physics. He carefully writes out all of its laws and lays down the initial positions and velocities of all the atoms in the universe. He then leaves to Saint Peter the tedious but intellectually trivial job of calculating all future happenings, including what, if any, macroscopic properties and macroscopic laws will emerge. That is the story of reductionism. Metaphysical pluralism supposes that God is instead very concerned about laws, and so he writes down each and every regularity that his universe will display. In this case Saint Peter is left with the gargantuan task of arranging the initial properties in the universe in some way that will allow all God's laws to be true together. The advantage to reductionism is that it makes Saint Peter's job easier. God may nevertheless have chosen to be a metaphysical pluralist.

6 Where inductions stop

I have argued that the laws of our contemporary science are, to the extent that they are true at all, at best true *ceteris paribus*. In the nicest cases we may treat them as claims about natures. But we have no grounds in our experience for taking our laws – even our most fundamental laws of physics – as universal. Indeed, I should say '*especially* our most fundamental laws of physics', if these are meant to be the laws of fundamental particles. For we have virtually no inductive reason for counting these laws as true of fundamental particles outside the laboratory setting – if they exist there at all. Ian Hacking is famous for the remark, 'So far as I'm concerned, if you can spray them then they are real.'[7] I have always agreed with that. But I would be more cautious: '*When* you can spray them, they are real.'

The claim that theoretical entities are created by the peculiar conditions and conventions of the laboratory is familiar from the social constructionists. The stable low-loss pulses I described earlier provide an example of how that can happen. Here I want to add a caution, not just about the existence of the theoretical entities outside the laboratory, but about their behaviour. Hacking's point is not only that when we can use theoretical entities in just the way we want to produce precise and subtle effects, they must exist; but also that it must be the case that we understand their behaviour very well if we are able to get them to do what we want. That argues, I believe, for the truth of some very concrete, context-constrained claims, the claims we use to describe their behaviour and control them. But in all these cases of precise control, we build our circumstances to fit our models. I repeat: that does not show that it must be possible to build a model to fit every circumstance.

Perhaps we feel that there could be no real difference between the one kind of circumstance and the other, and hence no principled reason for stopping our inductions at the walls of our laboratories. But there is a difference: some circumstances resemble the models we have; others do not. And it is just the point of scientific activity to build models that get in, under the cover of the laws in question, all and only those circumstances that the laws govern.[8] Fundamentalists want more. They want laws; they want true laws; but most of all, they want their favourite laws to be in force everywhere. I urge us to resist fundamentalism. Reality may well be just a patchwork of laws.

ACKNOWLEDGEMENTS

This chapter was originally published almost exactly as it appears here in Cartwright 1994.

[7] Hacking 1983, p. 23.

[8] Or, in a more empiricist formulation that I would prefer, 'that the laws accurately describe'.

2 Fables and models

1 New physics, new properties

Philosophers have tended to fall into two camps concerning scientific laws: either we are realists or we are instrumentalists. Instrumentalists, as we know, see scientific theories as tools, tools for the construction of precise and accurate predictions, or of explanations, or – to get down to a far more material level – tools for constructing devices that behave in the ways we want them to, like transistors, flash light batteries, or nuclear bombs. The laws of scientific theory have the surface structure of general claims. But they do not in fact make claims about the world; they just give you clues about how to manipulate it.

The scientific realist takes the opposite position. Laws not only appear to make claims about the world; they do make claims, and the claims are, for the most part, true. What they claim should happen is what does happen. This leads realists to postulate a lot of new properties in the world. Look at Maxwell's equations. These equations are supposed to describe the electromagnetic field: **B** is the magnetic intensity of the field and **E**, the electric intensity. The equations seem to make claims about the behaviour of these field quantities relative to the behaviour of other properties. We think that the equations are true just in case the quantities all take on the right values with respect to each other. There is thus a tendency, when a new theory is proposed, to secure the truth of its equations by filling up the world with new properties.

This tendency is nicely illustrated in some modern work on fibre bundles.[1] Nowadays we want our theories to be gauge-invariant. That implies that the Lagrangian should exhibit a local symmetry: $e^{i\lambda}\mathfrak{L}$ returns the same equations for motion as $\mathfrak{L}$ itself. (You may think of the Lagrangian as an elaborate way of representing the forces in a situation.) The λ here is a phase factor which used to be ignored; it was thought of, more or less, as an artefact of the notation. But now its behaviour must be regulated if the theory is to have the

[1] I have learned details of this work from J. B. Kennedy.

local symmetries we want. The values of λ are angles: 45°, 80°, 170°, and the like. This suggests constructing a space for the angle to rotate in. We can achieve the symmetries that gauge-invariance requires, but, as one author puts it, 'at the expense of introducing a new structure in our theory'. The structure is called a *principal fibre bundle*. It attaches an extra geometric space to each point in physical space. The primary motivation seems to be the practice of locating physical properties (e.g. field strengths) at one point in space. Here the property λ needs three dimensions to represent it. We end up with a space-time structure that has these fibres, or geometric 'balloons', attached to it at every point.

How shall we think about these structures? Most particle physicists prefer an instrumentalist reading of the geometry. But a minority take it realistically. The best example is probably Yural Ne'eman, who uses the structure to try to resolve standard paradoxes of quantum mechanics. The paradoxes are set in the famous paper by Einstein, Podolsky and Rosen. Ne'eman says:

> What makes [the phenomena of the] EPR [experiment] appear unreal is the fact that (with Einstein, Podolsky, and Rosen) we tend to visualize the experiment in the sectional setting of Galilean 3-space or even Riemannian 4-space, whereas the true arena is one in which the latter are just the base manifolds. It is in the additional dimensionality of the fibres . . . it is in this geometry that we should consider the measurement.[2]

Ne'eman is a wonderful example of the tendency to keep expanding the structures and properties of the universe as we expand our mathematical treatments of it.

It is this tendency that I want to resist. I want to defend the view that although the laws may be true ('literally' true), they need not introduce new properties into nature. The properties they mention are often already there; the new concepts just give a more abstract name to them. We tend to think of the concepts of physics, though theoretical, as very concrete, like *is red*; or *is travelling at 186,000 miles per second*. But it is better to take them as abstract descriptions, like *is a success*, or *is work*. In this case, I will argue, we have no need to look for a single concrete way in which all the cases that fall under the same predicate resemble each other. What we need to understand, in order to understand the way scientific laws fit the world, is the relationship of the abstract to the concrete; and to understand that, it will help to think about fables and their morals.

Fables transform the abstract into the concrete, and in so doing, I claim, they function like models in physics. The thesis I want to defend is that the

[2] Ne'eman 1986, p. 368.

relationship between the moral and the fable is like that between a scientific law and a model. Consider some familiar morals:

> It is dangerous to choose the wrong time for doing a thing.
> (Aesop, Fable 190)
>
> Familiarity breeds contempt.
> (Aesop, Fable 206)
>
> The weaker are always prey to the stronger.
> (G. E. Lessing)

Are these three claims true? They require in each case an implicit *ceteris paribus* clause since they describe only a single feature and its consequences; and they do not tell you what to expect when different features conflict. For example, when it comes to parents and children, the first claim is contradicted by the biblical moral, 'What man is there of you whom if his son ask bread will give him a stone?'[3] But the same is true of the laws of physics. In the case of picking up pins with a magnet, the law of gravity seems to be in conflict with the law of magnetic attraction. Both need at least to be read with an implicit *ceteris paribus* clause; or better, as a description of the nature of gravity or magnetic attraction. Barring these considerations, though, I would say that all three of these morals are most probably true. At least, there is no special problem about their truth. Nothing about their very nature prevents them from describing the world correctly.

So, too, with the laws of physics, I will argue. They can be true in just the same way as these very homely adages. But there is a catch. For making the parallel between laws and morals will allow us to limit the scope of scientific laws if we wish. Laws can be true, but not universal. We need not assume that they are at work everywhere, underlying and determining what is going on. If they apply only in very special circumstances, then perhaps they are true just where we see them operating so successfully – in the artificial environments of our laboratories, our high-tech firms, or our hospitals. I welcome this possible reduction in their dominion; but the fundamentalist will not.

2 The abstract and the concrete

In fact, I do not wish to make claims about what the correct relationship between the moral and the fable should be. Rather, I am interested in a particular theory of the relationship: that is the theory defended by Gotthold Ephraim Lessing, the great critic and dramatist of the German Enlightenment. His is a theory that sees the fable as a way of providing graspable, intuitive

[3] King James Bible, Matthew 7:9.

content for abstract, symbolic judgements. No matter whether Lessing has got it right or not about fables, it is his picture of the relationship between the moral as a purely abstract claim and the fable as its more concrete manifestation that mirrors what I think is going on in physics.

Lessing assigns three distinct roles to the fable. One is epistemological; one, ontological; and one, motivational. Epistemologically, Lessing follows Christian Wolff. Wolff distinguishes between *intuitive* (or *anschauende*) cognition and *figural*, which in the later aesthetic works of Mendelssohn, Maier and Lessing becomes the *symbolic*. In intuitive cognition we attend to our ideas of things; in figural or symbolic, 'to the signs we have substituted for them'. All universal and general knowledge, all systematic knowledge, requires symbolic cognition; yet only intuitive cognition can guarantee truth and certainty. For words are arbitrary, you cannot see through them to the idea behind them. Lessing's summarises the point in his essay on fables this way: 'Intuitive knowledge is clear in itself; the symbolic borrows its clarity from the intuitive'.[4] Lessing continues: 'In order to give to a general symbolic conclusion all the clarity of which it is capable, that is, in order to elucidate it as much as possible, we must reduce it to the particular in order to know it intuitively.'[5] That is the job of the fable. The general moral is a purely symbolic claim; the fable gives it a specific content so that we can establish with clarity the relation that is supposed to hold between the ideas. I shall say nothing further about this epistemological role, since it is not relevant to my claims about laws and models.

Clarity is not the only goal for Lessing. His primary interest is in motivation. He believes that intuitive recognition of truth has a far stronger influence on the will than mere symbolic recognition, and for this influence to be as strong as possible, you need to have a fully particular case in mind. This question of motivation is not relevant to my consideration about physics either, so I will not pursue it. What, rather, is crucial for me in Lessing's views is the claim that comes immediately between the two remarks above on intuitive cognition:

> The general exists only in the particular and can only become graphic (*anschauend*) in the particular.[6]

There are two clauses in this claim. The second is epistemological: 'the general only becomes graphic [or visualisable] in the particular'. The first is ontological: 'the general exists only in the particular'. It is this claim that matters here. To understand its significance, let us turn to the third moral on

[4] Lessing 1759 [1967], sec. I, p. 100.
[5] *Ibid.*
[6] *Ibid.*

the list above. The first two are from fables of Aesop. The third is a moral that goes in want of a fable. The fable is constructed by Lessing.[7]

> A marten eats the grouse;
> A fox throttles the marten; the tooth of the wolf, the fox.

Lessing makes up this story as a part of his argument to show that a fable is no allegory.[8] Allegories say not what their words seem to say, but rather something similar. But where is the allegory in the fable of the grouse, the marten, the fox and the wolf: 'What similarity here does the grouse have with the weakest, the marten with the weak, and so forth? Similarity! Does the fox merely *resemble* the strong and the wolf the strongest or *is* the former the strong, the latter the strongest. He *is* it.'[9] For Lessing, similarity is the wrong idea to focus on. The relationship between the moral and the fable is that of the general to the more specific, and it is 'a kind of misusage of the words to say that the special has a similarity with the general, the individual with its type, the type with its kind'.[10] Each particular is a case of the general under which it falls.

The point comes up again when Lessing protests against those who maintain that the moral is hidden in the fable, or at least disguised there. That is impossible given his view of the relationship between the two. Lessing argues: 'How one can *disguise* (*verkleiden*) the general in the particular, that I do not see at all. If one insists on a similar word here, it must at least be *einkleiden* rather than *verkleiden*.'[11] *Einkleiden* is to fit out, as when you take the children to the department store in the fall and buy them new sets of school clothes. So the moral is to be 'fitted out' by the fable.

The account of abstraction that I borrow from Lessing to describe how contemporary physics theories work provides us with two necessary conditions. First, a concept that is abstract relative to another more concrete set of descriptions never applies unless one of the more concrete descriptions also applies. These are the descriptions that can be used to 'fit out' the abstract description on any given occasion. Second, satisfying the associated concrete description that applies on a particular occasion is what satisfying the abstract description consists in on that occasion. Writing this chapter is what my working right now consists in; being located at a distance r from another charge q_2 is what it consists in for a particle of charge q_1 to be subject to the Coulomb force $q_1 q_2/4\pi\varepsilon_0 r^2$ in the usual cases when that force function applies. To say that working consists in a specific activity described with the relevant

[7] *Ibid.*, sec. I, p. 73.
[8] In fact he borrows the fable from von Hagedorn.
[9] *Ibid.*
[10] *Ibid.*
[11] *Ibid.*, sec. I, p. 86.

set of more concrete concepts on any given occasion implies at least that no further description using those concepts is required for it to be true that 'working' applies on that occasion, though surely the notion is richer than this.

Although I have introduced Lessing's account of the abstract and the concrete through a discussion of the fable and its moral, it marks an entirely commonplace feature of language. Most of what we say – and say truly – uses abstract concepts that want 'fitting out' in more concrete ways. Of course, that is compared to yet another level of discourse in terms of which they may be more concretely fitted out in turn. What did I do this morning? I worked. More specifically, I washed the dishes, then I wrote a grant proposal, and just before lunch I negotiated with the dean for a new position in our department. A well-known philosophical joke makes clear what is at stake: 'Yes, but when did you work?' It is true that I worked; but it is not true that I did four things in the morning rather than three. *Working* is a more abstract description of the same activities I have already described when I say that I washed dishes, wrote a proposal, and bargained with the dean.

This is not to say that the more abstract description does not tell you anything you did not already know from the more concrete list, or *vice versa*. They are, after all, different concepts. *Work* has implications about leisure, labour, preference, value, and the like, that are not already there in the description of my activity as washing the dishes or negotiating with the dean. (Though, I admit, I have chosen examples where the connection is fairly transparent for most of us.) Thus I am not suggesting any kind of reductionism of abstract concepts to more specific ones. The meaning of an abstract concept depends to a large extent on its relations to other equally abstract concepts and cannot be given exclusively in terms of the more concrete concepts that fit it out from occasion to occasion.

For the converse reason, the abstract–concrete relation is not the same as the traditional relation between genus and species. The species is defined in terms of the genus plus differentia. But in our examples, the more concrete cases, like washing dishes or travelling very fast, have senses of their own, independent (or nearly independent) of the abstractions they fall under. Nor should we think in terms of supervenience. Roughly, to say that one set of concepts supervenes on another is to say that any two situations that have the same description from the second set will also have the same description using the first set: the basic concepts 'fix' the values of those that supervene on them. This is not the case with the abstract–concrete distinction, as we can see from the example of *work*. Although washing dishes is what working amounted to for me in the early part of the morning, washing dishes only counts as working because certain other propositions using more abstract concepts like *preferences, leisure* and *value* are already presupposed for the situation. Should these fail, the very same activity need no longer count as work. Thus the notion of supervenience is in this sense stronger than the abstract–concrete relation described by Lessing.[12] The determinable–determinate relation is also stronger in just the same way.[13] For example, the determinable *colour* is fixed to hold as soon as any of the determinates that fall under it are fixed.[14]

Philosophical niceties aside, the important point about labelling *work* as an abstract concept in Lessing's sense is just that, in order to say truly that I worked, we do not have to assume that there was some other activity I did beyond those already mentioned. Consider the same point in the case of Lessing's fable, illustrated in figure 2.1. The marten is wily and quick; the grouse is slow and innocent. That is what it is for the grouse to be weaker than the marten. The fox is weaker than the wolf. But this is not a new relation between the fox and the wolf beyond the ones we already know so well and can readily identify in the picture: the wolf is bigger, stronger and has sharper teeth. That's what its being stronger than the fox consists in. It is just this ease of identification that accounts for the widespread use of animals in fables. Animals are used, Lessing maintains, because their characteristics are so well known and so permanent. About the use instead of particular persons, Lessing says, 'And how many persons are so generally well

[12] I have noticed that there is a tendency among reductionists of various kinds to try to collapse the distinction between abstraction and supervenience by arguing that in each case the entire abstract vocabulary will supervene on some more concrete description if only we expand the concrete descriptions to cover a broad enough piece of the surrounding circumstances ('global supervenience'). This is of course a metaphysical doctrine of just the kind I am disputing in this book.

[13] The determinable–determinate relation is stronger in a second way as well, since it requires that the designated determinate descriptions be mutually exclusive.

[14] This notion of supervenience – as well as Lessing's concept of abstraction – is also stronger than the notion of the abstract–concrete relation Jordi Cat has shown to be at work in Maxwell's discussions of concrete mechanical models *vis à vis* the more abstract descriptions in the energy-based Lagrangian formalism and its associated general principles of energy and work. The generality of the Lagrangian formalism, like that of a more 'abstract' phenomenological representation of electromagnetic phenomena in terms of electric and magnetic forces and energy (for Green, Maxwell and Heaviside), or that of the more 'general' representation of macroscopic media in continuum mechanics (for Stokes), lies in the elliptic assumption of the existence of an unknown underlying molecular structure represented by a mechanical model with hidden mechanisms – in which energy is manifested in motion (kinetic) or stored in elasticity (potential) – together with the realisation that an infinite number of more concrete mechanical descriptions can realise (or merely illustrate) the more abstract one. The more abstract one, however, needs independently to satisfy the mechanical principles that regulate and characterise the concepts of energy and force. See Cat 1995a, n. 23.

The supervenience relation is also, technically, weaker, for many definitions of supervenience do not formally require the first condition I take to be necessary for abstraction: to say that identical descriptions at the base level imply identical descriptions at the second level does not imply that no descriptions at the second level apply without some appropriate description from the base concepts, although this is often assumed.

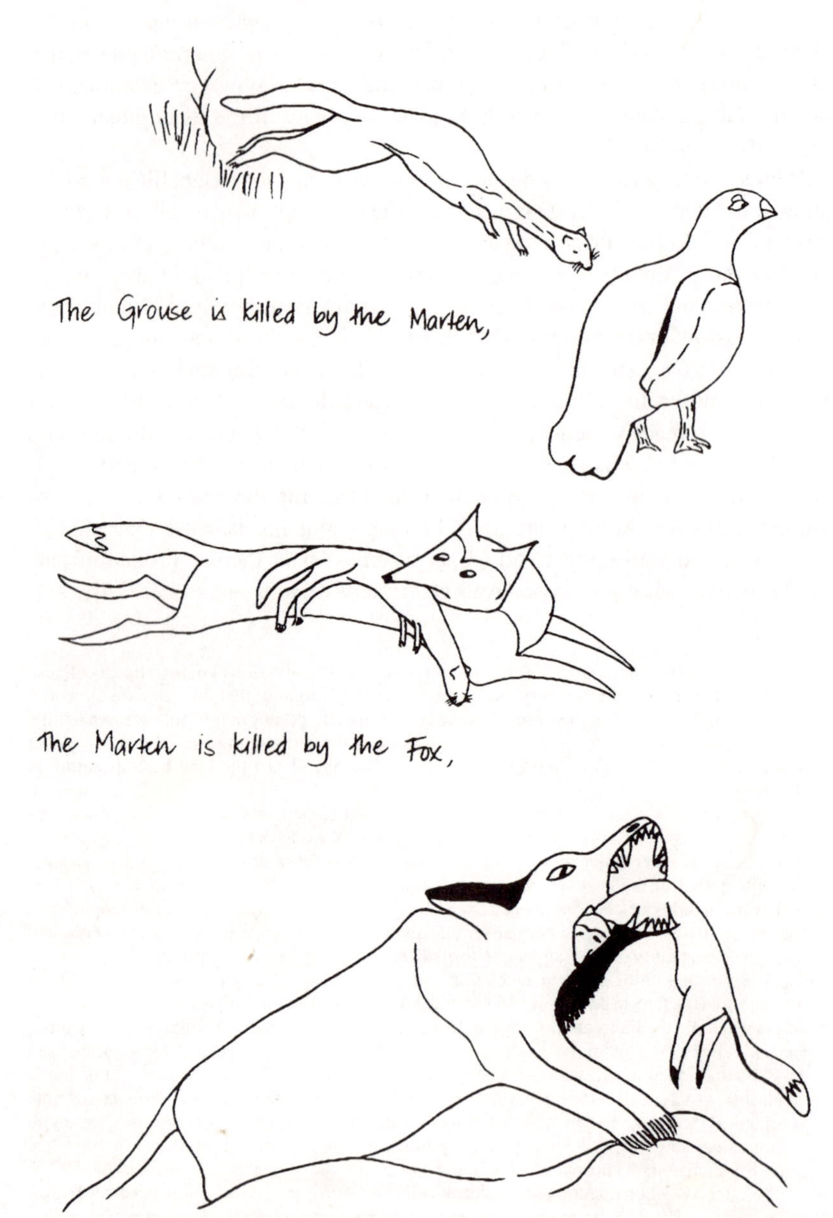

The Grouse is killed by the Marten,
The Marten is killed by the Fox,
And by the tooth of the wolf, the Fox.

Figure 2.1 Lessing's fable. Source: Rachel Hacking.

known in history that by merely naming them one can awake in everyone a concept of their corresponding way of thinking?'[15] For just this reason, though, *stereotypical* persons can serve. In *La Fontaine*, for example, the hero of the sour grapes tale is a Gascon – Gascon's typically being pictured as swaggering and boastful.

3 How physics concepts are abstract

Turn now from the Gascon and the fox to the stereotypical characters of the models which 'fit out' the laws of physics. Consider $\mathbf{F} = \boldsymbol{ma}$. I claim this is an abstract truth relative to claims about positions, motions, masses and extensions, in the same way that Lessing's moral 'The weaker are always prey to the stronger' is abstract relative to the more concrete descriptions which fit it out. To be subject to a force of a certain size, say **F**, is an abstract property, like *being weaker than*. Newton's law tells that whatever has this property has another, namely having a mass and an acceleration which, when multiplied together, give the already mentioned numerical value, **F**. That is like claiming that whoever is weaker will also be prey to the stronger.

In the fable Lessing proposes, the grouse is the stereotypical character exhibiting weakness; the wolf, exhibiting strength. According to Lessing we use animals like the grouse and the wolf because their characters are so well known. We only need to say their names to bring to mind what general features they have – boastfulness, weakness, stubbornness, pride, or the like. In physics it is more difficult. It is not generally well known what the stereotypical situations are in which various functional forms of the force are exhibited. That is what the working physicist has to figure out, and what the aspiring physicist has to learn.

This point can be illustrated by looking at the table of contents of a typical mechanics text, for example, the one by Robert Lindsay.[16] A major part of a book like this is aimed at the job I just described – teaching you which abstract force functions are exhibited in which stereotypical situations. That is like teaching you what everyone already knows about the grouse, that it is weak *vis-à-vis* the marten; or about the marten, that it is weak *vis-à-vis* the fox. Lindsay's text begins with a chapter of introduction, 'The Elemental Concepts of Mechanics'. Already the second chapter starts with simple arrangements, teaching us their force functions. Chapter Three continues with slightly more complicated models. Chapter Four introduces energy. Again though, immediately it turns to an account of what energy functions should be assigned to what situations: 'Energy Relations in a Central Force Field',

[15] *Ibid.*, sec. II, p. 110.
[16] Lindsay 1961.

'Inverse Square Field', 'Electron Energies in the Bohr Atom'. The same pattern is followed in the discussion of equilibrium (*e.g.*, 'Equilibrium of a Particle. Simple Cases . . . A System of Particles . . . Equilibrium of a Flexible String'), and similarly throughout the text.

Consider some simple examples of force functions. In the opening sections of Lindsay's chapter 2 we learn, for example, that the net force on a block pulled by a rope across a flat surface is given by $\mathbf{F}_r = \mathbf{F}_e - \mathbf{F}_f$, where $\mathbf{F}_e$ is from the pull of the hand, and $\mathbf{F}_f$ is due to friction. Under 'Motion in a Field Proportional to the First Power of the Distance', the basic model is that of a block on a spring. In this arrangement, $\mathbf{F} = -k\mathbf{x}$, where $\mathbf{x}$ is the displacement from the equilibrium position, and k is the spring constant. 'Motion in a Field Proportional to the Square of the Distance' is probably the most familiar since that is the form for the gravitational attraction between two masses. The simplest case is the two-body system. This is a situation in which a smaller body, m, is located a distance r from a larger, M. We learn that in this arrangement the small mass is subject to the force GmM/r^2.

Once we know how to pick the characters, we can construct a fable to 'fit out' Newton's law by leaving them alone to play out the behaviour dictated by their characters: that is what I called at the beginning, a *model* for a law of physics. For Lessing's moral, he picked the grouse and the marten. Now we can look to see if the grouse is prey to the marten. Similarly, we have the small mass m located a distance r from the larger mass M. Now we can look to see if the small mass moves with an acceleration GM/r^2. If it does, we have a model for Newton's law. Lessing said about his examples, 'I do not want to say that moral teaching is expressed (*ausgedrückt*) through the actions in the fable, but rather . . . that through the fable the general sentence is led back (*zurückgeführt*) to an individual case.'[17] In the two-body system, and similarly in each of the models listed in Lindsay's Table of Contents, Newton's law is 'led back' to the individual case.

Consider now Lessing's other claim, which I said at the beginning would be central to my argument:

> The general exists only in the particular, and can only become graphic in the particular.[18]

On my account *force* is to be regarded as an abstract concept. It exists only in the more specific forms to which it is led back via models of the kind listed in Lindsay's Table of Contents. It is not a new, separate property, different from any of the arrangements which exhibit it. In each case being in this arrangement – for example, being located at a distance r from another

[17] *Ibid.*, sec. I, p. 95.
[18] *Ibid.*, sec. I, p. 100.

massive body – is what it is to be subject to the appropriate force. *Force* is like my earlier example of *work* in this respect. I did not wash the dishes, apply for a grant, talk with the dean, and then work as well. These three activities constituted my working on the occasion. Similarly the world does not contain two-body systems with a distance *r* between them – or blocks and springs or blocks and rough surfaces – plus forces as well.

We can contrast this view with another which I think is more common, a view in which you can hive the force off from the situation and conceive it to exist altogether apart from the situation. Of course this is never thought to be actually possible since the situation is clearly physically responsible for the force; but it is taken to be logically possible. The simplest version of this idea of physical responsibility is that the situation produces the force (though it may not be by a standard causal process since perhaps the two come into existence together). The force comes into existence and it carries on, on its own; it is there in the situation in addition to the circumstances that produced it. This account makes *force* function like a highly concrete concept on the same level as the two bodies and their separation, or the block and the spring. It would be in this respect like a colour.

The barn is *red*. I painted it red. It could not have got to be red except through some causal process like that. Yet the redness is there above and beyond all those properties and processes which are responsible for its occurrence. There is, moreover, a natural sense of resemblance in which it is true to say that all red things resemble each other – the side of the barn I painted, the stripes on the American flag, a ripe tomato. When we model *force* on colour terms, we create the same expectation about it – all cases of forces must resemble each other. There must be something which 'looks' the same from one case to another. But obviously there is no simple way that these cases resemble each other, even across the three simple models I mentioned, let alone across all systems we want to describe as subject to forces. To achieve resemblance we need to invent a new property – the separate and logically independent property of being-subject-to-a-force. Admittedly, it is not really a 'visualisable' property – we cannot figure out what it looks like. Still, we have the confidence that whatever its characteristics are, they resemble each other from one occasion to the next.

By now it is clear what I think of this story. *Force* – and various other abstract terms from physics as well – is not a concrete term in the way that a colour predicate is. It is, rather, abstract, on the model of *working*, or *being weaker than*; and to say that it is abstract is to point out that it always piggy-backs on more concrete descriptions. In the case of *force*, the more concrete descriptions are ones that use the traditional mechanical concepts, such as *position, extension, motion*, and *mass*. *Force* then, on my account, is abstract relative to these concepts from mechanics; and being abstract, it can only exist in particular mechanical models. In chapter 8 I will provide a number of further examples of how abstract terms get made concrete in physics and how this constrains the scope of application for these terms.

4 Truth and social construction

I conclude with some possible lessons about truth, objectivity and realism in physics. Nowadays the social constructivists provide us with powerful arguments against taking the laws of physics as mirrors of nature. Scientists, after all, operate in a social group like any other; and what they do and what they say are affected by personal motives, professional rivalries, political pressures, and the like. They have no special lenses that allow them to see through to the structure of nature. Nor have they a special connection or a special ear that reveals to them directly the language in which the Book of Nature is written. The concepts and structures that they use to describe the world must be derived from the ideas and concepts that they find around them. We can improve these concepts, refine the structure, turn them upside down, inside out; we can even make a rather dramatic break. But always the source must be the books of human authors and not the original Book of Nature. What we end up with through this process is bound to be a thoroughly human and social construction, not a replica of the very laws that God wrote.

What then of the startling successes of science in remaking the world around us? Don't these successes argue that the laws on which the enterprise is based must be true? Social constructivists are quick to point out that the successes are rather severely restricted to just the domain I mentioned – the world as we have made it, not the world as we have found it. With a few notable exceptions, such as the planetary systems, our most beautiful and exact applications of the laws of physics are all within the entirely artificial and precisely constrained environment of the modern laboratory. That in a sense is a commonplace. Consider one of the founders of econometrics, Tyrgve Haavelmo, who won the Nobel prize in 1989 for his work in originating this field. Haavelmo remarks that physicists are very clever. They confine their predictions to the outcomes of their experiments. They do not try to predict the course of a rock in the mountains and trace the development of the avalanche. It is only the crazy econometrician who tries to do that, he says.[19]

Even when the physicists do come to grips with the larger world they do not behave like the econometrician, argue the social constructivists. They do not take laws they have established in the laboratory and try to apply them outside. Rather, they take the whole laboratory outside, in miniature. They

[19] Personal conversation, 1992.

construct small constrained environments totally under their control. They then wrap them in very thick coats so that nothing can disturb the order within; and it is these closed capsules that science inserts, one inside another, into the world at large to bring about the remarkable effects we are all so impressed by. The flashlight battery is a good example.

Another illustration is provided by the measurement example mentioned in the last chapter of the magnetic field gradients for neuromagnetism in order to look for evidence of strokes. The authors of a paper on the SQUIDS employed in such measurements explain:

> One measures minute magnetic signals . . . emanating from the electrical currents in neurons in the brain. These signals must be measured in the presence of both man-made and natural magnetic field noise that can be [far higher]. An elegant way of rejecting the ambient interference in favor of the locally generated signal is by means of a gradiometer . . . To reduce magnetic field fluctuations even further, it is becoming common practice to *place the instruments and patient in a shielded room consisting of multiple layers of mu-metal and aluminium.*[20]

The conclusion I am inclined to draw from this is that, for the most part, the laws of physics are true only of what we make. The social constructivists tend to be scornful of the 'true' part. There is almost always the suggestion lurking in their writings that it is no surprise that the laws work for the very situations they have been designed to work for. The scientists in turn tend to shrug their shoulders in exasperation: 'You try for a while and you'll find out. It is a major achievement to get anything to work, and it is just as hard to get a good model to describe it when it does.'

The observation that laws of physics are general claims, like the morals of fables, and that the concepts they employ are abstract and symbolic can provide a middle ground in the dispute. Newton's law, for instance, can be true of exactly those systems that it treats successfully; for we have seen how we can take it to be true of any situation that can be simulated by one of the models where the force puts on a concrete dress. That does not mean that we have to assume that Newton has discovered a fundamental structure that governs all of nature. That is part of the point of seeing force as an abstract concept, like work, and not a more concrete one, like extension. We may ask of any object, 'What is its extension?' and expect there to be an answer. But not work. My children's teachers used to say to me 'Play is a child's work'. Were they right? I am inclined to think this is one of those situations where they were neither right or wrong. The normal activities of middle-class preschoolers are not an arena to which we can lead back the abstract concept of *work* and its relations, like *labour, value* and *leisure*. The concepts do not apply here.

[20] Clarke and Koch 1988, p. 220, italics added.

Similarly with force. You may assume that for every object – this book, a ship in the Atlantic Ocean, or the rocks sliding over each other in the avalanche – there is an answer to the question, 'What is the force on this object?' But you need not. Whether you do will depend on how widely you think our models apply. I have argued that the laws are true in the models, perhaps literally and precisely true, just as morals are true in their corresponding fables. But how much of the world are fables true of? I am inclined to think that even where the scientific models fit, they do not fit very exactly. This question bears on how true the theory is of the world. But it is a different question from the one at stake here. Choose any level of fit. Can we be assured that for every new situation, a model of our theory will fit at that level, whether it be a model we already have, or a new one we are willing to admit into our theory in a principled way? This is a question that bears, not on the truth of the laws, but rather on their universality.

What about force? You may agree with the great British physicist Kelvin[21] that the Newtonian models of finite numbers of point masses, rigid rods, and springs, in general of inextendable, unbendable, stiff things can never simulate very much of the soft, continuous, elastic, and friction-full world around us. But that does not stop you from admitting that a crowbar is rigid, and, being rigid, is rightly described by Newton's laws; or that the solar system is composed of a small number of compact masses, and, being so composed, it too is subject to Newton's laws. It is a different question to ask, 'Do Newton's laws govern all of matter?' from 'Are Newton's laws true?' Once we recognise the concept of force as an abstract concept, we can take different views about how much of the world can be simulated by the models that give a concrete context to the concept. Perhaps Newton's models do simulate primarily what we make with a few fortuitous naturally occurring systems like the planets to boot. Nevertheless, they may be as unproblematically true as the unexceptionable and depressing claim that the weaker are prey to the stronger.

ACKNOWLEDGEMENTS

This chapter was originally published in almost exactly the form it appears here in Cartwright 1991. My thanks to Conrad Wiedeman for his help and to Stanford-in-Berlin for the opportunity of continuing my studies of Lessing; also to J. B. Kennedy for discussions of Ne'eman.

[1] I have learned about Kelvin from Norton Wise. *Cf.* Smith and Wise 1989.

3 Nomological machines and the laws they produce

1 Where do laws of nature come from?

Where do laws of nature come from? This will seem a queer question to a post-logical-positivist empiricist. Laws of nature are basic. Other things come from, happen on account of, them. I follow Rom Harré[1] in rejecting this story. It is capacities that are basic, and laws of nature obtain – to the extent that they do obtain – on account of the capacities; or more explicitly, on account of the repeated operation of a system of components with stable capacities in particularly fortunate circumstances. Sometimes the arrangement of the components and the setting are appropriate for a law to occur naturally, as in the planetary system; more often they are engineered by us, as in a laboratory experiment. But in any case, it takes what I call a *nomological machine* to get a law of nature.

Here, by *law of nature* I mean what has been generally meant by 'law' in the liberalised Humean empiricism of most post-logical-positivist philosophy of science: a law of nature is a necessary regular association between properties antecedently regarded as OK. The association maybe either 100 per cent – in which case the law is deterministic, or, as in quantum mechanics, only probabilistic. Empiricists differ about what properties they take to be OK; the usual favourites are sensible properties, measurable properties and occurrent properties. My objections do not depend on which choice is made. The starting point for my view is the observation that no matter how we choose our OK properties, the kinds of associations required are hard to come by, and the cases where we feel most secure about them tend to be just the cases where we understand the arrangement of capacities that gives rise to them. The point is that our knowledge about those capacities and how they operate in given circumstances is not itself a catalogue of modalised regularity claims. It follows as a corollary from my doctrine about where laws of nature come from that laws of nature (in this necessary regular association

[1] *Cf.* Harré 1993.

sense of 'law') hold only *ceteris paribus* – they hold only relative to the successful repeated operation of a nomological machine.

What is a nomological machine? It is a fixed (enough) arrangement of components, or factors, with stable (enough) capacities that in the right sort of stable (enough) environment will, with repeated operation, give rise to the kind of regular behaviour that we represent in our scientific laws. The next four chapters will argue for the role of nomological machines in generating a variety of different kinds of laws: the laws we test in physics, causal laws, results in economics and probabilistic laws. This chapter aims to provide a sense of what a nomological machine is and of why the principles we use to construct nomological machines or to explain their operation can not adequately be rendered as laws in the necessary regular association sense of 'law'.

2 An illustration from physics of a nomological machine[2]

Consider the naturally occurring regularity I mentioned above: planetary motion. Kepler noted that Mars follows an elliptical orbit with the sun at one focus. This is described in the first law that bears his name. Since the time of Robert Hooke and Isaac Newton, the so-called 'Kepler's problem' has been just how to account for such an observed regularity in terms of mechanical descriptions, that is, using descriptions referring to material bodies, their states of motion and the forces that could change them. Specifically the account calls for the mechanical description of the system in terms of an arrangement of two bodies and their connection. In my terminology the task is to figure out the nomological machine that is responsible for Kepler's laws – with the added assumption that the operation of the machine depends entirely on mechanical features and their capacities. This means that we have to establish the arrangement and capacities of mechanical elements and the right shielding conditions that keep the machine running properly so that it gives rise to the Kepler regularities.

The basic insight for how to do so was shared not only by Hooke and Newton but also by their successors well into our century, for instance, Richard Feynman.[3] If a stone attached to a string is whirling around in a circle, it takes a force to modify its direction of motion away from the straight path so as to keep it in the circle. What is required is a pull on the string. In abstract physical terms, it takes a radial attractive force – a force along the

[2] I want to thank Jordi Cat for discussions and for contributing significantly to this section of the chapter.

[3] Feynman 1992.

radius of the circular path and towards its centre. In the case of the orbiting planet, the constituents of the nomological machine are the sun, characterised as a point-mass of magnitude M, and the planet, a point-mass of magnitude m, orbiting at a distance r and connected to the former by a constant attractive force directed towards it. Newton's achievement was to establish the magnitude of the force required to keep a planet in an elliptical orbit:

$$F = -GmM/r^2$$

where G is the gravitational constant. The shielding condition is crucial here. As we well know, to ensure the elliptical orbit, the two bodies must interact in the absence of any additional massive body and of any other factors that can disturb the motion.

Newton solved Kepler's problem by showing that the elliptical geometry of the orbit determines the inverse-square kind of attraction involved in the gravitational pull.[4] Conversely, he also showed that an attraction of that kind in the circumstances described could give rise to the observed regularity of the elliptical motion of Mars.[5] Although his proofs were essentially geometrical in character, the well-known equivalent analytical proofs were soon introduced and adopted.[6] In both cases, built into the mechanical concept of *force* is the assumption that in the right circumstances a force has the capacity to change the state of motion of a massive body. To account for the regular motion of the different planets in the solar system, we must describe a different machine, with a modified arrangement of parts and different shielding conditions. The pull of the additional planets on any planet whose observed regular orbit is being considered is then added to the pull exerted by the sun

[4] See I. Newton, *Principia*, Proposition 11. The proof solves the so-called 'direct Kepler's problem'. See Newton 1729.

[5] *Ibid.*, Proposition 13. The proof provides a solution to the so-called 'inverse Kepler's problem'.

[6] A sketch of the analytical proof goes as follows. In the relation between a force, the mass of a body and the acceleration the body undergoes as a result of the force (alternatively, the acceleration by virtue of which it is able to exert the force – Newton's second law of motion – the force can be expressed as a function of the position of the body and of time: $\mathbf{F}(\mathbf{x},t) = m\, d^2\mathbf{x}/dt^2$. A transformation into polar co-ordinates, the radius r and the angle ϕ, allows for an expression of the force in terms of the radial co-ordinates, $F(r,t)$ and the angular ones, $F(\phi,t)$. By eliminating the time parameter, one can obtain an expression for the force in terms of r and ϕ only: $F(r,\phi) = -l^2/mr^2(d^2(1/r)/d\phi^2 + 1/r)$, where l is the angular momentum of the system. This is the 'polar orbital equation'. Then by differentiating the orbital equation of the ellipse, $1/r = c(1 + e\cos\phi)$ (e is the eccentricity and c is a constant), one can arrive at the inverse-square form of the required force, $F = -k'(1/r^2)$, in the direction of the sun, where k' is empirically determined by the arrangement of the system ($k' = GmM$). A discussion of Newton's geometrical proofs and their correspondence with the modern analytical substitutes can be found in Brackenridge 1995.

in the expression of the gravitational force. The resulting orbits typically deviate from Kepler's perfect ellipses.[7]

The example of the planetary motions is important for me since it has been used by philosophers and physicists alike in support of the view that holds more 'basic' regularities as first and fundamental in accounting for observed regularities (i.e., in explanation, it is laws 'all the way down'). This view emphasises the unifying power of the appeal to Newton's laws with respect to Kepler's. I do not deny the unifying power of the principles of physics. But I do deny that these principles can generally be reconstructed as regularity laws. If one wants to see their unifying power, they are far better rendered as claims about capacities, capacities that can be assembled and reassembled in different nomological machines, unending in their variety, to give rise to different laws. Newton's 'law of gravitation' is not a statement of a regular association between some occurrent properties – say masses, distances and motions. For it does not tell us about the motions of two masses separated by a distance r, but, instead, about the *force* between them. The term 'force' in the equation of gravity does not refer to yet another occurrent property like mass or distance that could appear in a typical philosopher's list of occurrent properties. Rather, it is an abstract term (in the sense of chapter 2) that describes the capacity of one body to move another towards it, a capacity that can be used in different settings to produce a variety of different kinds of motions.

Those who advocate laws as fundamental also point to the heuristic role they play in scientific discovery. Thus Feynman writes, '[W]hen a law is right it can be used to find another one. If we have confidence in a law [e.g., Newton's law of gravitation], then if something appears to be wrong it can suggest to us another phenomenon.'[8] Feynman is referring to the discovery of Neptune. The belief in Neptune's existence was suggested by the irregularity that the orbit of Uranus displayed with respect to the predictions that can be made from Newtonian principles. In the law-first view, this discovery speaks to the importance of universal laws. I think this claim is mistaken. The observed irregularity points instead to a failure of description of the specific circumstances that characterise the Newtonian planetary machine. The discovery of Neptune results from a revision of the shielding

[7] It is worth mentioning that Newton's nomological machine derives its unifying power from its ability to account in addition for the regularities described in Kepler's second and third law. But in the case of the third law (that the square of the period of a planet's motion is proportional to the cube of the major axis of its orbit) Newton's description of the setting includes the assumption that the planet's mass is negligible compared to the mass of the sun. Kepler's law describes then only an approximation to the actual regularities displayed by the larger planets, such as Jupiter and Saturn. *Cf.* Goldstein 1980, p. 101.

[8] Feynman 1992, p. 23.

conditions that are necessary to ensure the stability of the original Newtonian machine.[9]

3 Models as blueprints for nomological machines

My views about nomological machines come primarily from my work on models in the LSE Modelling and Measurement in Physics and Economics Project. When we attend to the workings of the mathematical sciences, like physics and economics, we find the important role models play in our accounts of what happens; and when we study these models carefully we find that they provide precisely the kind of information I identify in my characterisation of a nomological machine. Let us consider first models that lie entirely within a single exact science, such as physics or economics, where the role of the word 'exact' is to point to the demand made on models in these disciplines that the behaviour to be explained within the model should be rigorously derived from facts about the model plus the principles of the theory. I consider a number of different models in various chapters here. On the physics side these include Newton's planetary models for Kepler's laws, which I just discussed; the detailed model provided by the Stanford Gravity-Probe team for the predicted precession of the four gyroscopes they are sending into space, from chapter 4; and the BCS model for the regular behaviour described in London's equations, from chapter 8. There are in addition two extended examples from economics: one in chapter 6 describes the regularities about efficiencies and inefficiencies that arise when debt contracts are written in certain ways and the other, an association between the length of time that individuals are unemployed and the persistence of unemployment in the economy, from chapter 7. All of these models provide us with a set of components and their arrangement. The theory tells us how the capacities are exercised together

In order to do this job, the capacities deployed in the models we construct in the exact sciences will differ from the more ordinary capacities we refer to in everyday life. Consider, for example, Coulomb's law. Coulomb's law describes a capacity that a body has *qua* charged. It differs from many everyday ascriptions of capacities in at least three ways that are important to the kind of understanding that exact science can provide of how a nomological machine operates. First, the capacity is associated with a specific feature –

[9] Of course, the alleged universality of the capacity of two masses to attract one another as described in Newton's principles does matter for that is the usual justification for the assumption that *these* particular masses will attract each other. Weaker assumptions about the extent of the capacity claim would clearly serve as well if the assumption of true universality should seem too grand.

charge – which can be ascribed to a body for a variety of reasons independent of its display of the capacity described in the related law – here Coulomb's law. This is part of what constitutes having a scientific understanding of the capacity. Contrast two more everyday ascriptions. I am *irritable* and my husband is *inaccurate*. These are undoubtedly capacities we have. Ask the children or anyone we work with. Each has been established on a hundred different occasions in a hundred different ways. Like the Coulomb capacity, these too are highly generic. They give rise to a great variety of different kinds of behaviour; the best description of what they share in common is that they are displays of my irritability or Stuart's inaccuracy.

These everyday cases contrast with the scientific examples that I am concerned with in the ways we have available to judge when the capacity obtains and when it does not. No one claims in cases like irritability to point to features which you could identify in some other way, independent of my displays of irritability, that would allow you to determine that I am indeed irritable. Philosophers debate about whether there must be any such features: first, whether there need be any at all in the individual who has the capacity, and second how systematic must be the association between the features and the capacity across individuals. Whatever the answer to these questions about everyday capacities, part of the job in science is to find what systematic connections there are and to devise a teachable method for representing them.

The second way that Coulomb's capacity differs from everyday ones is that it has an exact functional form and a precise strength, which are recorded in its own special law. Third, we know some very explicit rules for how the Coulomb capacity will combine with others described by different force laws to affect the motions of charged particles. What happens when a number of different forces are exerted together on the same object? To find out, we are taught to calculate a 'total' force ($\mathbf{F}_t$) by vector addition of the strengths and directions recorded in each of the related force laws separately. Then we use the formula $\mathbf{F}_t = \boldsymbol{ma}$ to compute the resulting acceleration.

These two features are characteristic of the study of capacities in exact science, although the method of representation varies significantly across domains, both for the capacities themselves and for how to calculate what happens when they operate jointly. For an example of a different method of representation in physics we can move from the study of bodies in motion to that of electric circuits. The capacities of the components of a circuit – resistors, capacitors, inductances, and impedances – are represented in well-known formulae. For instance, the *capacitance* of an isolated conductor is $C = Q/V$. How do we calculate the current in a complex circuit from knowledge of the capacities of its components? We reduce the complex circuit to a simpler equivalent one that has elementary well-known behaviour, using some selection from a vast variety of circuit-reduction theorems, such as

Thévenin's theorem or Millman's theorem. Then we use what is essentially Ohm's law ($I = V/R$) to calculate the current.

In game theory various concepts of equilibrium describe what is supposed to happen when the capacities of different agents are all deployed at once. Since we will look in some detail at an example from game theory in chapter 6, here let us turn to the more descriptive side of economics, to econometrics. In econometrics strengths of capacities are generally represented by coefficients of elasticity, which can often be measured by partial conditional expectations. There are two different standard ways for calculating what happens under combination. One is similar to vector addition. We add together the canonical influences from each feature separately. So we end up with a linear equation:[10] effects are represented by the independent variable, different causes by the dependent variables, where the respective coefficients represent the 'strength' of the capacity of each separate cause.[11]

In different situations we proceed differently: we use a set of simultaneous equations to fix what happens when different capacities are exercised together. Each separate capacity is represented by a different equation. When a number of capacities are exercised together, all the equations must be satisfied at once. Consider the simple case of supply and demand. The capacity of price to affect quantity supplied is represented in an upward sloping line; its capacity to affect quantity demanded by a line that slopes downward.

$$q_s = \alpha p + \mu \qquad \alpha > 0$$

$$q_d = \beta p + v \qquad \beta < 0$$

If the system is in equilibrium, the quantity supplied equals the quantity demanded:

$$q_s = q_d$$

What happens when both capacities of the price are exercised together? We require then that all the equations be satisfied at once. This means that the price is fixed; it lies at the intersection of the supply and demand curves. This is the source of the well-known identification problem in economics: how do we identify the equations we should use to represent the supply and demand equations separately when the supply mechanism and the demand mechanism never work on their own? What happens is always far more limited than what either equation allows, since the patterns of behaviour permitted by one are always further constrained by the second. The problem is

[10] It is of course possible to introduce more complicated functional forms, as we generally do in physics. One cost is that most of the statistical techniques we usually employ for testing will no longer be available.

[11] These kinds of cases were considered at length in Cartwright 1989.

especially pressing here because it does not even make sense to think of either of the two capacities being exercised on its own.[12]

These examples bring out the wholistic nature of the project we undertake in theory formation in exact science. We must develop on the one hand *concepts* (like 'the force due to gravity', 'the force due to charge'[13] or 'capacitance', 'resistance', 'impedance' . . .) and on the other, *rules for combination*; and what we assume about each constrains the other, for in the end the two must work together in a regular way.[14] When the concepts are instantiated in the arrangements covered by the rules, the rules must tell us what happens, where *regularity* is built into the demand for a rule: *whenever* the arrangement is thus-and-so, what happens is what the rule says should happen.

Developing concepts for which we can also get rules that will work properly in tandem with them is extremely difficult, though we have succeeded in a number of subject areas. In both physics and economics we have a variety of formal theories with special concepts and explicit rules that allow us to predict what regular behaviours should occur whenever the concepts are instantiated in the prescribed kinds of arrangements. And in physics, where we have been able to build clear samples of these arrangements, a number of our formal theories are well confirmed. Economics generally must rely on a more indirect form of testing, and the verdicts there are far less clear. At any rate, the success in various branches of physics in devising special concepts and laws that work in cases where the concepts clearly apply shows that there are at least some domains where the requirements we have been discussing are not impossible to fulfil.

A common metaphysical assumption about the completeness (or completability) of theory would go further and put an even more severe demand on our scientific concepts. The assumption was well expressed by John Stuart Mill:

> The universe, so far as known to us, is so constructed that whatever is true in any one case is true in all cases of a certain description: the only difficulty is to find what description.[15]

The sense of *completeness* I have in mind is this: a theory is complete with respect to a set of cases when it supplies for those cases the descriptions that Mill expects plus the principles that connect the descriptions.

[12] This is in sharp contrast with the method of representation just discussed in which factors with different capacities are combined in a single equation. Generally in these cases the value of the other relevant causes can be set to 'zero' to represent situations in which they do not operate.

[13] In my vocabulary these would be called 'the Coulomb capacity', 'the capacity for gravitational attraction' and so on.

[14] This point, I take it, is similar to that of Donald Davidson in Davidson 1995.

[15] Mill 1843, vol. 1, p. 337.

What about this additional demand? Should we accept it? I urge 'no'. The constraints imposed on concept formation in exact science by the demands to build at the same time a system of matching rules that will work together with the concepts in the right way are so severely confining that we have only satisfied them in a few formal theories in physics, and with great effort. And even in physics, we never have had a success, nor a near success, at completeness. It is only subject to the big *ceteris paribus* condition of the operation of an appropriate nomological machine that we can ever expect, 'that whatever is true in any one case is true in all cases'. We might well of course aim for completeness in any case where we have an empirically well-grounded research programme that offers promising ideas for how to achieve it. But in general we have no good empirical reason to think the world at large lends itself to description by complete theories.

This is why the idea of a nomological machine is so important. It is, after all, only a philosophical concept, like 'unconditional law' or 'complete theory' or 'universal determinism', a way of categorising and understanding what happens in the world. But it has the advantage over these that it adds less than they do to what we are given in our observations of how successful formal theories work; and it shows that we do not need to use these more metaphysically extensive concepts in order to make sense of either the successes of our exact sciences nor of the pockets of precise order that these sciences can describe. Where there is a nomological machine, there is law-like behaviour. But we need parts described by special concepts before we can build a nomological machine. The everyday concepts of *irritability* and *inaccuracy* will not do, it seems, nor the concept of acceleration in terms of rate of change of velocity with distance ($d\mathbf{v}/dx$) rather than with time ($d\mathbf{v}/dt$), which the Medievals struggled to make a science of. We also need a special arrangement: a bunch of resistors and capacitors collected together in a paper bag will not conduct an electric current. When we understand it like this, we are not inclined to think that exact science must be completable, at least in principle, in order to be possible at all.

There is one further central aspect of nomological machines that I have so far not discussed: *shielding*. Recall the irregularity in the orbit of Uranus from the point of view of the original model of the planetary machine. This reminds us that is not enough to insist that the machine have the right parts in the right arrangement; in addition there had better be nothing else happening that inhibits the machine from operating as prescribed. As we saw in chapter 1, even a very basic principle like $\mathbf{F} = m\boldsymbol{a}$ needs a shield before it can describe a regularity. We can have all the forces in all the right arrangements that license assignment of a particular 'total' force $\mathbf{F}$. But we cannot expect an acceleration $\boldsymbol{a} = \mathbf{F}/m$ to appear if the wind is blowing too hard. The need for shielding is characteristic of the ordinary machines we build in everyday life. The importance of the concept of shielding in understanding when regularities arise is a large part of the reason why it is so useful to think of the special arrangements that generate regularities as *machines*.

Models of certain kinds, I claim, function as blueprints for nomological machines. But we must not mistake this for the claim that the models we usually see in theories show us to build a nomological machine. The models are generally given at far too high a level of abstraction for that. Just think about the arrangements that must obtain in the model when we expect to do a vector addition of a number of forces represented there: the forces must all be 'exercised together'. And what does that mean? At least in the case of certain descriptive concepts, we get help from the bridge principles of the theory. But we generally get no advice at all in the case of arrangements. Even the bridge principles of course are little help in the actual building of a machine. Bridge principles tell us what the abstract concepts consist in more concretely. (For examples see chapter 8.) But what we are told is still too formal. We need to know about real materials and their properties, what the abstract concepts amount to *there*, before we can build anything. But telling us this is no part of theory. This is one of the reasons that I find the 'vending machine' view of scientific prediction, testing, or application that I discuss in chapter 8 so grotesque.

I began with models that lie entirely within a single exact science. These are the models that allow us to predict in a systematic and warranted way the kind of precise and regular behaviour that we see in the laboratory and in many of our carefully manufactured technological devices or even occasionally in nature as it comes. But not all regular behaviour is precise. Coarsely-tuned machines, like my old bicycle, can give us regular behaviour even though the descriptions under which the behaviour falls are in no way quantitatively precise. Nor are the predictions of models always warranted in this top-down way. In general we construct models with concepts from a variety of different disciplines, the arrangements in them do not fit any rules for composition we have anywhere and the regular behaviour depicted in the model does not follow rigorously from any theory we know. Yet these models too, whenever I look at one of them, seem well described as blueprints for nomological machines.

So here is my strong claim: look at any case where there is a regularity in the world (whether natural or constructed) that we judge to be highly reliable and which we feel that we understand – we can either explain the regularity or we believe it does not need explanation. What you will find, I predict, is that the explanation provides what is clearly reasonable to label as a *nomological machine*. And where there is no explanation needed you will still find a machine. Sometimes for instance the whole situation is treated as one simple machine (like the lever), where the shielding conditions and the idea of repeated operation are so transparent that they go unnoted. To the extent that this claim is borne out, to that extent we have powerful empirical evidence that you cannot get a regularity without a nomological machine. And if nomological machines are as rare as they seem to be, not much of what happens in nature is regular and orderly, as Mill supposed it to be. The world is after all deeply dappled.

4 Capacities: openness and invention

I argue against laws that are unconditional and unrestricted in scope. Laws need nomological machines to generate them, and hold only on condition that the machines run properly. But there are, as we saw in the last section, some very well understood machines, modelled within the various disciplinary boundaries of our exact sciences. I say our understanding of these depends on knowledge of capacities, not knowledge of laws. Is there much, after all, in the difference? I think so, because when we refuse to reconstruct our knowledge as knowledge of capacities, we deny much of what we know and we turn many of our best inventions into pure guesses. What is important about capacities is their open-endedness: what we know about them suggests strategies rather than underwriting conclusions, as a vending-machine view of science would require. To see the open-endedness it is useful to understand how capacities differ from dispositions.

Disposition terms, as they are usually understood, are tied one-to-one to law-like regularities. But capacities, as I use the term, are not restricted to any single kind of manifestation. Objects with a given capacity can behave very differently in different circumstances. Consider Coulomb's law, $F = -q_1q_2/4\pi\varepsilon_0r^2$, for two particles of charge q_1 and q_2 separated by a distance r. I will discuss this case in more detail in chapter 4. For here let us just consider what Coulomb's law tells us about the motions of the particle pair. It tells us absolutely nothing. Before any motion at all is fixed, the particles must be placed in a special kind of environment; just the kind of environment that I have described as a nomological machine. Without a specific environment, no motion at all is determined.

We may think that the *natural* behaviour for opposite charges is to move towards each other and for similar charges, to separate from each other. But it is important to keep in mind that this is not an effect *in abstracto*. That motion, like any other, depends on how the environment is structured. There is no one fact of the matter about what *occurs* when charges interact. With the right kind of structure we can get virtually any motion at all. We can even create environments in which the Coulomb *repulsion* between two negatively charged particles *causes them to move closer together*. Figure 3.1 gives an

Two electrons e_1 and e_2 are released from rest into a cylinder as in Figure 3.1b. The cylinder is open from one side only, and it is open to a unified magnetic field directed towards the negative z-axis. The initial distance between the two electrons is r_1. According to the laws of electromagnetism, the force between the two electrons is a repulsive force equal to

$$F = \frac{1}{4\pi\varepsilon_0}\frac{e_1 e_2}{r_1^2} = m_e a_a.$$

Whereas e_2 will be locked inside the cylinder, e_1 will enter the magnetic field **B** with a certain velocity v_1. The magnetic field on e_1 will move it in a circular motion (as in the figure) with a force equal to

$$F = ev_1 \otimes B.$$

This will take the electron e_1 into an insulated chamber attached to the cylinder. The dimensions of the cylinder and the chamber can be set so that the distance between the final position of e_1 and e_2 is less than r_1.

Figure 3.1a Source: example constructed by Towfic Shomar.

example, due to Towfic Shomar, from the LSE Modelling and Measurement in Physics and Economics Project.

For a different kind of example, let us turn to economics, to a study by Harold Hotelling[16] of Edgeworth's taxation paradox.[17] This is a case that I have worked on with Julian Reiss, also from the LSE Modelling and Measurement in Physics and Economics Project.[18] Taxes have the capacity to affect prices. How do we characterise the effects of this capacity? Think again about the capacity represented in Coulomb's law with respect to the motion of oppositely charged particles. We tend to characterise this capacity in the canonical terms I used above: opposite charges move towards each other; similar charges, away from each other. Similarly, *taxes increase prices*. The 'paradox' pointed out by Edgeworth is that this is not the only possibility. In the right situations taxes can decrease prices, and they can do so by following just the same principles of operation that 'normally' lead to price increase.

Hotelling produced a toy model of a simple economy that illustrates Edgeworth's paradox. The economy consists of many firms which compete in the production of the different commodities and many buyers whose sole source of utility derives from these goods. A version of the Hotelling economy with

[16] Hotelling 1932.
[17] *Cf.* Edgeworth 1925, section II.
[18] See also Hands and Mirowski 1997 and my comments in Cartwright 1997c.

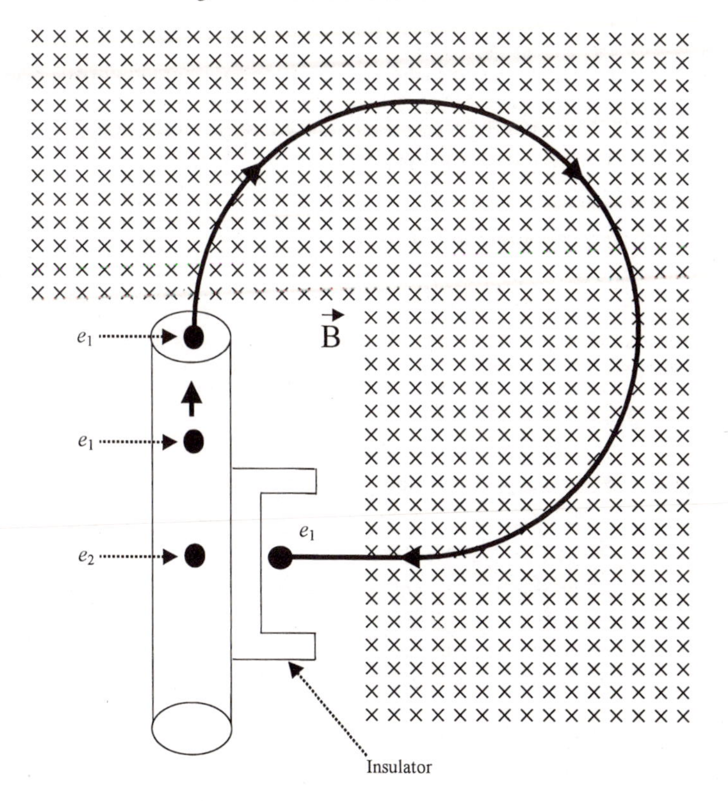

Figure 3.1b.

only two goods is described in Figure 3.2. If we consider a tax *t* levied on the first good only, what matters are two equations of the following forms:

$$dp_1 = tA/D$$

$$dp_2 = tB/D$$

A and *B* are functions, through the demand and supply equations, of partial derivatives with respect to each of the two quantities of functions of both quantities $(\delta f(q_1 q_2)/\delta q_1, \ldots)$. In the terminology I used in *Nature's Capacities and their Measurement*,[19] *A*/*D* represents *the strength of t's capacity to affect dp_1*, and *B*/*D*, *the strength of t's capacity to affect dp_2*. For Hotelling's two commodity economy, it can be shown that *D* is always positive. But it is possible to construct supply and demand functions whose parameters make

[19] Cartwright 1989.

The economy[1] consists of many firms who compete in the production of two different commodities and many buyers whose sole source of utility is from these two goods. The prices of the goods are p_1 and p_2, respectively, and the demand functions are given by the expressions:

$$q_i = F_i(p_1, p_2) \quad (i = 1, 2). \tag{1}$$

It is assumed that these equations are solvable, such that we have the inverse demand functions:

$$p_i = f_i(q_1, q_2) \quad (i = 1, 2). \tag{2}$$

For the producers the respective supply functions are given with the following expressions:

$$q_i = G_i(p_1, p_2) \quad (i = 1, 2) \tag{3}$$

$$p_i = g_i(q_1, q_2) \quad (i = 1, 2). \tag{4}$$

Now, let $h_i(q_1, q_2)$ be the excess of demand price over supply price. Thus we obtain,

$$h_i = f_i - g_i. \tag{5}$$

A differentiation with respect to a q_i will be denoted with a subscript j, so that,

$$f_{ij} \equiv \frac{\partial f_i}{\partial q_j}, \quad g_{ij} \equiv \frac{\partial g_i}{\partial q_j}, \quad h_{ij} \equiv \frac{\partial h_i}{\partial q_j}.$$

The last definition will be the determinant of the marginal excess price matrix:

$$D \equiv \frac{\partial(h_1, h_2)}{\partial(q_1, q_2)} = \begin{vmatrix} h_{11} & h_{12} \\ h_{21} & h_{22} \end{vmatrix} = h_{11}h_{22} - h_{21}h_{12}.$$

Let an asterisk * denote the equilibrium values for which supply and demand are equal. Then

$$h_i(q_1^*, q_2^*) = 0. \tag{6}$$

Now, a tax t_i per unit sold is imposed on the ith commodity, payable by the producers. Let $p_i + \mathrm{d}p_i$ and $q_i + \mathrm{d}q_i$ be the new prices and quantities. In equilibrium, the demand price must exceed the supply price by exactly t_i. Hence,

$$h_i(q_1^* + \mathrm{d}q_1, q_2^* + \mathrm{d}q_2) = t_i.no(7)$$

A Taylor expansion of the first order and the subtraction of equation (6) yield the following approximations for small t_i:

$$h_{i1}\mathrm{d}q_1 + h_{i2}\mathrm{d}q_2 = t_i. \tag{8}$$

Figure 3.2 Taxation under free competition in a Hotelling economy. Source: construction of this two-commodity economy is by Julian Reiss.

The solutions to these equations are:

$$dq_1 = \frac{1}{D}\begin{vmatrix} t_1 & h_{12} \\ t_2 & h_{22} \end{vmatrix} = \frac{t_1 h_{22} - t_2 h_{12}}{h_{11}h_{22} - h_{21}h_{12}},$$

$$dq_2 = \frac{1}{D}\begin{vmatrix} h_{11} & t_1 \\ h_{21} & t_2 \end{vmatrix} = \frac{t_2 h_{11} - t_1 h_{21}}{h_{11}h_{22} - h_{21}h_{12}}. \qquad (9)$$

The interesting effect the tax has is on the prices. The price changes to buyers resulting from the taxes are:

$$dp_i = -\frac{1}{D}\begin{vmatrix} 0 & f_{i1} & f_{i2} \\ t_1 & h_{11} & h_{12} \\ t_2 & h_{21} & h_{22} \end{vmatrix}. \qquad (10)$$

Or, more specifically, the h_{ij}s replaced by the corresponding $f_{ij} - g_{ij}$ and, following Hotelling's (1932) example, the tax levied upon good one only ($t_1 = t$, $t_2 = 0$):

$$dp_1 = \frac{t(f_{11}f_{22} - f_{11}g_{22} - f_{12}f_{21} + f_{12}g_{21})}{D}$$

$$dp_2 = \frac{t(f_{22}g_{21} - f_{21}g_{22})}{D}. \qquad (11)$$

Edgeworth's taxation paradox arises when a tax levied upon a good decreases its price rather than increases it, as is normally expected. It can be shown that for two commodities $D > 0$.[2] With that result and the fact that $t > 0$ the conditions for Edgeworth taxation paradox to arise are:[3]

$$dp_1 < 0 \Leftrightarrow f_{11}f_{22} - f_{12}f_{21} < f_{11}g_{22} - f_{12}g_{21},$$

$$dp_2 < 0 \Leftrightarrow f_{21}g_{22} - f_{22}g_{21} < 0. \qquad (12)$$

One can easily see the dual (or triple) capacity of the tax to increase or decrease (or leave unchanged) the prices, depending on the values of the other parameters. This dual capacity stems from the fact that the two goods interact both in consumption and production. For a single commodity equation (10) yields:

$$dp = -\frac{1}{D}\begin{vmatrix} 0 & f_{11} \\ t & h_{11} \end{vmatrix} = \frac{tf_{11}}{h_{11}} > 0, \qquad (13)$$

since t is always positive and both f_{11} and h_{11} $(= f_{11} - g_{11})$ are negative.

[1] This example is a simplified version of the model of Hotelling (1932), Section 5.
[2] *Cf.* Hotelling (1932), p. 601.
[3] The slight differences between these conditions and Hotelling's original conditions (25) and (26) arise from the fact that Hotelling makes use of his integrability conditions that imply $h_{ij} = h_{ji}$.

Figure 3.2 cont.

A or *B* or both negative. So, depending on the specific structure of supply and demand in the Hotelling economy, taxes can increase prices, or decrease them, or the prices may even stay the same. Yet in all instances it is the same capacity at work, with the same functional form, derived from the same basic principles.

These two examples – of Coulomb's law and of the capacity of taxes to affect prices – illustrate why I talk of *capacities*, which give rise to highly varied behaviours, rather than of *dispositions*, which are usually tied to single manifestations. If we do wish to stick to the more traditional vocabulary of dispositions, then a capacity is what in *The Concept of Mind* Gilbert Ryle called a 'highly generic' or 'determinable' disposition as opposed to those that are 'highly specific' or 'determinate'. According to Ryle, verbs for reporting highly generic dispositions 'are apt to differ from the verbs with which we name the dispositions, while the episodic verbs corresponding to the highly specific dispositional verbs are apt to be the same. A baker can be described as baking now, but a grocer is not described as ''grocing'' now, but only as selling sugar now, or weighing tea now, or wrapping up butter now.'[20]

The point I want to stress is that capacities are not to be identified with any particular manifestations. They are rather like 'know', 'believe', 'aspire', 'clever' or 'humorous' in Ryle's account: 'They signify abilities, tendencies, propensities to do, not things of one unique kind, but things of lots of different kinds.'[21] This is why the idea of the nomological machine is so important when we think of using the knowledge we gather in our exact sciences to intervene in the world. Much of modern scientific theory is about capacities, capacities which can have endless manifestations of endless different varieties. That is the key to how scientific invention is possible. Similarly charged particles repel each other, opposite charges attract; what that can amount to in terms of the motions and locations of the particles is limited only by our imagination. Taxes affect prices, but what happens to the prices depends on the economics we build and how well we build them.

5 Do we really need capacities?

It is now time to defend explicitly my claim that we need claims about capacities to understand nomological machines and cannot make do with laws, in the necessary regular association sense of 'law'. I shall look at two prominent places where we can see why we need capacities instead of laws. The first is in the principles for building nomological machines, the second for describing

[20] Ryle 1949, p. 118.
[21] *Ibid.*, p. 119.

their running. It is important to the discussion to keep firmly in mind that there is more to the conventional sense of 'law' than regularity: the regularities are supposed to be ones between some especially favoured set of OK properties, say occurrent properties or ones we can measure. But these will never give us what we need.

Look first to the building of a nomological machine. In building the machine we compose causes to produce the targeted effect. Consider again Newton's principle of gravity and Coulomb's law. These two may work together, in tandem with Newton's second law of motion (**F** = ***ma***), to explain the trajectory of a charged body. I say that Newton's and Coulomb's principles describe the capacities to be moved and to produce motion that a charged body has, in the first case the capacity it has on account of its gravitational mass and in the second, on account of its charge. How should we render Newton's principle, instead, as a claim about regular associations among purely occurrent or directly measurable properties?

The relevant vocabulary of occurrent or measurable properties in this case is the vocabulary of motions – positions, speeds, accelerations, directions and the like. But there is nothing in this vocabulary that we can say about what masses do to one another.[22] As we saw in section 4, when one mass attracts another, it is completely open what motion occurs. Depending on the circumstances in which they are situated, the second mass may sit still, it may move towards the first, it may even in the right circumstances move away. There is no one fact of the matter about what occurrent properties obtain when masses interact. But that does not mean that there is no one thing we can say. 'Masses attract each other.' That is what we say, that is what we test, in thousands of different ways; and that is what we use to understand the motions of objects in an endless variety of circumstances.

'Masses attract each other.' Perhaps this is what regularity theorists had in mind all along. But if so, they have given up the resistance to capacity talk altogether. What occurrent or directly measurable properties do two bodies have in common when they are both attracted to another body? None. Similarly two masses that are both busy attracting other bodies are crucially alike, but not in any way that can be described within the vocabulary of measurable or occurrent properties. Think about Gilbert Ryle's arguments in *The Concept of Mind*. When we use the term 'attract' in the consequent of a regularity claim, we do just what Ryle warns us against in the case of mental dispositions: we categorise together as one kind of episode all the results that happen

[22] The one case we have looked at where the basic principles could legitimately be thought of as describing what systems do using only occurrent-property language is in the simultaneous equations models of econometrics. The equations are supposed to involve only measurable quantities, and since each equation must be separately satisfied, the relations between measurable quantities that really occur are literally in accord with each of the principles.

when two masses interact, whatsoever these episodes look like. The *Concise Oxford English Dictionary*,[23] for instance, defines 'attract' when used 'of a magnet, gravity, etc.' as 'exert a pull on'. 'Attract' and 'pull' are like 'groce' for the activities of a grocer and 'solicit' for the activities of a solicitor. They are not in the usual philosopher's list of occurrent property terms. Rather, they mark the fact that the relevant capacity has been exercised. That is what is in common among all the cases when masses interact as Newton described.

Sometimes we conceal the widespread use in physics of terms like 'attract', terms that mark the exercise of a capacity, by a kind of equivocation. We switch back and forth between an occurrent sense of the term – a body has *attracted* a second when the second moves towards it – in which Newton's principle or Coulomb's is generally not borne out – and the sense marking the exercise of a capacity in which the principles do seem to be true (if not universally at least as widely as we have looked so far). 'Attract', like many verbs in both ordinary and technical language, comes with a natural effect attached, and with two senses. In the first sense the natural effect must occur if the verb is to be satisfied; in the second sense, it is enough for the system to exercise its capacity regardless of what results, i.e., for it *to try to produce the associated effect*.

The trying is essential, and sometimes verbs like these have it built right into their definition. To 'court', according to the *Concise Oxford Dictionary*,[24] is to 'try to win the affection or favour of (a person)'. These kinds of words are common in describing the facts of everyday life: to brake – to apply the brakes, or to succeed in slowing the vehicle; to anchor – to lower the anchor, or to succeed in securing the boat; to push, to pull, to resist, to retard, to damn, to lure, to beckon, to shove, to harden (as in steel), to light (as the fire), . . .; and especially for philosophers: to 'explain' is not only, in its first sense in the *Concise Oxford Dictionary*, to 'make . . . intelligible', but also, in its second, to 'say *by way of* explanation'.[25]

The technical language of physics shares this feature with our more ordinary language; indeed it shares much of the same vocabulary. *Attraction, repulsion, resistance, pressure, stress*, and so on: these are concepts that are essential to physics in explaining and predicting the quantities and qualities we can directly measure. Physics does not differ from ordinary language by needing only some special set of occurrent property terms or directly measurable quantities stripped of all connections with powers and dispositions. Rather, as I described in section 3, what is distinct about the exact sciences

[23] 8th edn, 1990.
[24] *Ibid.*
[25] *Ibid.*, italics added.

is that they deal with capacities that can be exercised not only more or less – push harder, or resist less – but with ones for which the strength of effort is quantifiable, and for which, in certain very special circumstances, the exact results of the effort may be predictable.

The second place where it is easy to see the need for capacity concepts is when we set the nomological machine running. This is a point I make much of at other places, so I will only summarise here. Consider the very simple case of two charged bodies separated by a distance *r*. To calculate their motions, we add vectorially the force written down in Coulomb's principle and the force written down in Newton's law of gravity; then we substitute the result into Newton's second law, $\mathbf{F} = m\boldsymbol{a}$. What then are we supposing? First, that there is nothing that inhibits either object from exerting both its Coulomb and its gravitational force on the other; second, no other forces are exerted on either body; and third, everything that happens to the two bodies that can affect their motions can be represented as a force. Notice that these caveats all have to do with capacities and their exercise. Nothing must inhibit either the charges or the gravitational masses from exercising their capacities. No further capacities studied in classical dynamics should be successfully exercised; and finally, the capacity of a force to move a body as recorded in Newton's second law must be exercised successfully, unimpeded and without interference.

Can we render these caveats without using the family of concepts involving capacities? Throughout these chapters I argue that we cannot. (In particular, I treat the first and second conditions in chapters 4 and 8; the third was discussed at length in chapter 1.) The idea that we can do so is part of the fundamentalist pretensions of physics: there is some vocabulary special to physics within which we can describe everything that matters to the motions of bodies. This view gains support, I take it, from a mistaken understanding about how deductivity works in physics. In theories like mechanics, electromagnetism and special relativity we have had considerable success in finding sets of occurrent property descriptions that have a kind of deductive closure: certain kinds of effects describable in that vocabulary occur reliably in circumstances where all the causes of these kinds of effects (and their arrangement) can be appropriately described within the designated vocabulary. But that does not cash out into regularity laws with the descriptions of the causes and their arrangements in the antecedents and the descriptions of the effects in the consequent. For we still need the shielding: nothing else must occur that interferes with the capacities of those causes in that arrangement to produce those effects.

The need for this kind of addition is often obscured by the plasticity of the language of physics. Sometimes terms in physics refer to genuinely measurable quantities that objects or systems might possess and sometimes the use of the very same terms requires truths about the operation of capacities for its satisfaction. This plasticity gives physics two different ways to finesse the problems I have been discussing – either narrow the range of the antecedent to include the *ceteris paribus* conditions right in it, or expand the range of the consequent to cover whatever occurs when the capacities in question operate.

We have seen lots of illustrations of the second device already, for instance with the introduction of words like 'attract' and 'repel' into Coulomb's law and the law of gravity. The first can be seen in the simple case of the law of the lever. 'Lever' can be defined in terms of occurrent properties, making no allusion to capacities and their exercise. So, we sometimes use 'lever' to mean *rigid rod*, where a rod is *rigid* just in case the distances between all the mass points that make it up remain constant through the motions of these mass points. But sometimes we use 'lever' only for rigid rods so placed that their capacity to exhibit the behaviour required in the law of the lever will operate unimpeded. In this sense (if physics is right about the capacities of rigid rods), then a lever is bound to satisfy the law of the lever.

So far I have argued that there are jobs we do – and indeed should do – with our scientific principles that cannot be done if we render them as laws instead of as descriptions of capacities. There is one answer to my plea for capacities that sidesteps these defences of capacities. The answer employs a kind of transcendental argument. It does not attempt to show how it is possible to do these jobs without capacities but rather tries to establish that it must be possible to do so. I borrow the form from arguments made by Bas van Fraassen and by Arthur Fine in debating more general questions of scientific realism.[26] The argument presupposes that we have available a pure data base, cleansed of capacities and their non-Humean relativities. The objection goes like this: 'You, Cartwright, will defend the design of given machine by talking about what impedes and what facilitates the expression of the capacities in question. I take it this is not idle faith but that in each case you will have reason for that judgement. These reasons must ultimately be based not in facts about capacities, which you cannot observe, but in facts about actual behaviour, which you can. Once you have told me these reasons, I should be able to avoid the digression through capacities and move directly to the same conclusions you draw with capacities. Talk of capacities may provide a convenient way to encode information about behaviours, but so long as we insist that scientific claims be grounded in what can be observed, this talk cannot contribute any new information.'

But what about this decontaminated data base? Where is it in our experience? It is a philosophical construction, a piece of metaphysics, a way to

[26] Van Fraassen 1980, Fine 1986.

interpret the world. Of course we cannot do without interpretation. But this construction is far more removed from our everyday experience of the world as we interact with it and describe it to others than are homely truths about triggering mechanisms, precipitating factors, impediments, and the like, which mark out the domain of capacities. Consider an adaptation of van Fraassen's objection to causes,[27] which is a version of essentially the same argument. The objection proceeds from the assumption that there is some defensible notion of a sensible property which is conceptually and logically distinct from any ideas connected with capacities. We are then confronted with a challenge to explain what difference capacities make. 'Imagine a world identical with our own in all occurrences of its sensible qualities throughout its history but lacking in facts about capacities. How would that world differ from our world?'

On one reading, this argument may be about sequences not of properties in the world but of our experiences about the world. These sequences are to remain the same, but we are to imagine that they are not caused in the usual way by what is going on in the world around us. This reading cannot be the one intended, though, since it does not cut in the right way, revealing special virtues for descriptions like 'is red' or 'is a jet-stream trail' in contrast with ones like 'has the power to relieve headaches' or 'attracts other charges, *qua* charged'.

I might further be invited to inspect my experiences and to notice that they are 'really' experiences of succession of colour patches, say, with capacities nowhere to be found. The philosophical dialogue along this line is well rehearsed; I merely point in the familiar directions. My experiences are of people and houses and pinchings and aspirins, all things which I understand, in large part, in terms of their capacities. I do not have any raw experience of a house as a patchwork of colours. Even with respect to colours, my experience is of properties like *red*, which brings to objects the capacity to look specific ways in specific circumstances. Sense data, or *the given*, are metaphysical constructs which, unlike capacities, play no role in testable scientific claims. Once there was a hope to mark out among experiences some raw pieces by using an epistemological yardstick: the 'real' experiences were the infallible ones. After a great deal of debate it is not clear whether this criterion even lets in claims about felt pains; but it surely does not distinguish claims like 'The stripes are red' from 'Your pinching makes my arm hurt' and 'Mama is irritable'.

The contemporary version of this argument tends, for these reasons, not to be in terms of sense experiences but in terms of sensible properties. But here there is a very simple reply. A world with all the same sensible properties as

[27] Van Fraassen 1980, ch. 5.

ours would already be a world with capacities. As I remarked above, redness is the property that, among other things, brings with it capacity to look just *this* way in normal circumstances, and to look systematically different when the circumstances are systematically varied.

Perhaps we are misled here by carrying over the conclusions of an earlier metaphysics, conclusions for which the premises have been discarded. These premisses involve the doctrine of impressions and ideas. In the immediately post-Cartesian philosophy of the British empiricists, sensible properties could be picked out because they looked like their impressions. Gaze at the first stripe on the American flag: redness is the property that looks like *that*. We do not have this copy theory; so we do not have properties that are identified like that. Correlatively, we can no longer make the same distinction separating powers and their properties as did these seventeenth-century empiricists. On their doctrine, the way things looked could get copied in the perceiver's impressions of them; but the various powers of the property could not. Since their ideas were copies of their impressions, necessarily their world, as imaged, had only inert properties.

But we do not have the copy theory of impressions, nor do we adopt this simple theory of concept formation. For us, there are properties, and all properties bring capacities with them. (Perhaps, following Sydney Shoemaker,[28] they are all just conglomerates of powers.) What they are is given not by how they look but by what they do. So, 'How does the Hume world differ from ours?' It would not differ. Any world with the same properties as ours would *ipso facto* have capacities in it, since what a property empowers an object to do is part of what it is to be that property. The answer is the same for a world with the same sensible properties. And what about a world the same with respect to all the look-of-things? That question may have made sense for Locke, Berkeley, and Hume; but without the copy theory of impressions and the related associationist theory of concept formation, nowadays it makes no sense.

6 Metaphysical aside: what makes capacity claims true?

The world is made up of facts, Wittgenstein taught us. As empiricists we should insist that it is these facts that make our scientific claims true. What

[28] Shoemaker 1984, ch. 10.

facts then are they that make our capacity claims true? Let me turn the question around and see what the traditional view has to say. What facts make law claims true in the necessary regular association sense of law? There are, I think, only two honest kinds of answer for an empiricist.

The first is that *regularities make law claims true*, real regularities, ones that actually occur. These are undeniably facts in the world, not for instance putative facts in some merely possible world. They are thus proper empiricist candidates for truth makers. But we know that this lets in both too little and too much. Start with too much. What about all the accidental regularities? There is an honest empiricist answer:[29] laws are those regularities that cover the widest range of occurrences in the most efficient way.[30] The objection that there are too few regularities was taken up by Bertrand Russell:[31] a good many of the claims we are most interested in, especially in contexts of forecasting and planning, are about situations that may never occur or only rarely get repeated. Russell claimed that physics solves this problem by using very abstract descriptions; at that level 'the same thing' does generally occur repeatedly. (So, for example, the trajectories of the planets and of cannon balls and of electrons in a cloud chamber are all supposed to instantiate $\mathbf{F} = \boldsymbol{ma}$.)

My objection is the same in both cases. I summarise the lessons argued for in various places throughout this book: there are no such regularities to begin with. Unless we take capacities robustly, Coulomb's and Newton's principle are ruled out immediately. Perhaps they are to get relegated to the status of calculational tools for getting 'real' regularities, like $\mathbf{F} = \boldsymbol{ma}$. But even this is not a true regularity without adding to the antecedent the caveat that the force *operates unimpeded*. Russell's proposal fares better; but, as I argued in section 5, only if we allow our abstract vocabulary to include terms like 'attract' and 'repel', terms that have implications about capacities and their operations built in. So regularity theorists cannot even get started unless they too take facts involving how capacities operate to be part of the constitution of the world.[32]

Alternatively, there are proposals[33] to take necessitation as one of the kinds of facts that make up the world. Then we can still be *empiricist* in the sense that we can stick to the demand that scientific claims be judged against facts

29 *Cf.* Friedman, Earman, Kitcher.
30 That is, what makes a law claim true are first, a regular association and second, facts about how much a given collection of regular associations covers versus how much another does.
31 Russell 1912–13.
32 Facts, for instance, of the form *X interfered with Y's capacity to do Z.*
33 *Cf.* Maudlin 1997.

about the real world around us.[34] The drawback to this proposal from my point of view is not that it lets modal facts into the world but rather that it lets in the wrong kind of modal fact. The inversion of a population of atoms does not *necessitate* the emission of coherent radiation; it allows it. But it allows it in some very special way. After all, anything *can* cause anything else. In fact, it seems to me not implausible to think that, with the right kind of nomological machine, almost anything can *necessitate* anything else. That is, you give me a component with a special feature and a desired outcome, and I will design you a machine where the first is followed by the second with total reliability. Just consider, for example, Rube Goldberg machines or Paolozzi sculptures.

So, if anything can cause practically anything else, what is special about the claim that an inversion in a population of atoms allows (or can cause) coherent radiation? We can use the expression we often introduce in explaining our intuitions about laws of nature here: the inversion allows the coherent radiation *by virtue of the structure of the world*, or *by virtue of the way the world is made*. But what does that mean? To mark the distinction between the kind of accidental possibility, where anything can result in anything else, and this other more nomological sense of possibility, Max Weber labelled the latter '*objective* possibility'.[35] Weber's ideas seem to me very much worth pursuing in our contemporary attempts to understand scientific knowledge. But so far I still think that the best worked out account that suits our needs most closely is Aristotle's doctrines on *natures*, which I shall defend in the next chapter. Capacity claims, about charge, say, are made true by facts about what it is in the nature of an object to do by virtue of being charged. To take this stance of course is to make a radical departure from the usual empiricist view about what kinds of facts there are.

That returns me to the plea for the scientific attitude. Philosophical arguments for the usual empiricist view about what there is and what there is not are not very compelling to begin with. They surely will need to be given up if they land us with a world that makes meaningless much of what we do and say when we use our sciences most successfully. What makes capacity claims true are facts about capacities, where probably nature's grammar for capacities is much like our own – or at least as much like our own as any

[34] Perhaps I should say that this allows us to satisfy the *ontological* demands of empiricism. There are of course in addition in the empiricist canon also epistemological demands and demands about how meanings can be fixed. In my view, as I argue in different places here and in Cartwright 1989, all these kinds of demands are just as well met by claims about capacities as by claims about occurrent properties.

[35] *Cf.* Weber's *The Logic of Historical Explanation* in Runciman 1978, which is translated from Weber's 1951. In that essay, Weber attributes the concept of *objective possibility* to the German physiologist Johannes von Kries.

other claims about the structure of the world that we back-read from successful scientific formulations. What makes true, then, the claim, 'Inversion in a population of atoms has the capacity to produce coherent radiation'? In simple Tarski style, just that: the fact that inversion has the capacity to produce coherent radiation. And this fact, so far as our evidence warrants,[36] has as much openness about it with respect to determining occurrent properties, as does our own claim about the capacity.

7 Nomological machines and the limits of science

I have been defending the claim that facts about capacities and how they operate are as much a part of the world as pictured by the exact sciences as are facts about occurrent properties and measurable quantities. One may be inclined to query what all the fuss is about. Once we have forsaken the impressions-and-ideas theory of concept formation defended by Hume and all forms of sense-data theories as well, how are we to draw a distinction between facts about occurrent properties and ones about capacities in the first place?

I have no qualms about giving up the distinction. But in doing so we must not lose sight of one important feature of capacities that affects our doctrines about the limits of science. There is no fact of the matter about what a system can do just by virtue of having a given capacity. What it does depends on its setting, and the kinds of settings necessary for it to produce systematic and predictable results are very exceptional. I have argued here that it takes a nomological machine to get a regularity. But nomological machines have very special structures. They require the conditions to be just right for a system to exercise its capacities in a repeatable way, and the empirical indications suggest that these kinds of conditions are rare. No matter how much knowledge we might come to have about particular situations, predictability in the world as it comes is not the norm but the exception. So we should expect regularities to be few and far between. If we want situations to be predictable, we had better engineer them carefully.

ACKNOWLEDGEMENTS

This chapter is dedicated to Rom Harré, who taught me that it is all right to believe in powers. I wish to thank Jordi Cat, who has contributed substantially to section 2

[36] Again I should add that I do not think there are any successful arguments that the evidence here is less good than for any of the more usual empiricist claims about what kinds of facts there are. Indeed, if my arguments are right, the evidence is much better since a reconstruction of scientific claims using capacity language will go very much farther in capturing our empirical successes than will a reconstruction that uses only the language of occurrent properties.

and has provided valuable editorial help and useful conversations, and Towfic Shomar and Julian Reiss who developed the examples in section 4. Work on this chapter was supported by the LSE Modelling and Measurement in Physics and Economics Project. About half of this chapter has been previously published. Sections 1 and 2, bits of section 5 and all of section 7 appeared in much the same form in Cartwright 1997b. Section 3 is entirely new as are the crucial examples of section 4 and all of section 6. Some of section 5 appeared in Cartwright 1992.

Part II

Laws and their limits

The laws we test in physics
Causal laws
Current economic theory
Probabilistic laws

4 Aristotelian natures and the modern experimental method

1 Beyond regularity

Law-like regularities result from the successful operation of nomological machines. But what kinds of facts (if any) determine the behaviour of a nomological machine? The Humean tradition, which finds nothing in nature except what regularly happens, insists that it must be further regularities. This chapter will continue the argument that laws in the sense of claims about what regularly happens are not our most basic kind of scientific knowledge. More basic is knowledge about capacities, in particular about what capacities are associated with what features. (Or, put more laboriously, 'knowledge about what capacities a system will have in virtue of having particular features'.) 'More basic' here is meant neither in an epistemic nor in an ontological sense. Epistemically, capacity knowledge and knowledge of regularities are on an equal footing – neither is infallible and both are required if we are to learn anything new about the other. And ontologically, even though reliable regularities will not come about unless capacities are harnessed appropriately, nevertheless claims about how things regularly behave in given circumstances are neither more nor less true nor more nor less necessary than claims about the capacities that explain why things behave regularly in these circumstances. Rather knowledge of capacities is more basic in that it is both more embracing and more widely useful than knowledge of regularities.

Virtue is not all on the side of capacities though. To affect the world around us we need to make reliable predictions about it and that requires regularities. What can be relied on to happen in a given situation is what regularly happens in it, or would regularly happen were enough repetitions to occur. Knowledge of the capacities that various features of the world bring with them is not enough to tell us what reliably happens in any situation at all. We need only recall our nomological machines to see that. We need very special knowledge about how capacities can be harnessed before we can expect any regularities.

As I have urged in chapter 3, knowledge of capacities is a quite different kind of knowledge than knowledge of what things do or what they can do, of what happens or what can happen. John Stuart Mill talked of knowledge like this as knowledge of what things *tend* to do. But that kind of talk is rarely heard now.[1] Our contemporary account of scientific knowledge is dominated instead by categories favoured by British empiricists from the eighteenth and the twentieth centuries, most notably Locke, Berkeley, Hume, Ayer and Ryle. These are the categories of the 'sensible' or 'observable' properties and of properties that are 'occurrent' as opposed to powers and dispositions. But we get a far better description of scientific knowledge if we adopt a category of Aristotle's. The knowledge we have of the capacity of a feature is not knowledge of what things with that feature do but rather knowledge of the *nature* of the feature. This chapter will explain why I reject the conventional categories of British empiricism and turn instead to more ancient ones. A concept like Aristotle's notion of *nature* is far more suitable than the concepts of *law, regularity* and *occurrent property* to describe the kind of knowledge we have in modern science: knowledge that provides us the understanding and the power to change the regularities around us and produce the laws we want.

2 Historical background

One of the great achievements of the scientific revolution, according to its adherents, was the banishment from modern science of the Aristotelian schemes of explanation which had dominated Scholastic studies. Aristotle was derided as a cuttlefish, a squid: the ink he discharged cast everything into obscurity. Consider one typical case, Pierre Gassendi in his *Exercises against the Aristotelians*.[2] Gassendi complains that Aristotelian explanations fail to explain. About the definition of motion as 'the act of being in potentiality insofar as it is in potentiality', he remarks: 'Great God! Is there any stomach strong enough to digest that? The explanation of a rather familiar thing was requested, but this is so complicated that nothing is clear anymore . . . The need for definitions of the words in the definitions will go on *ad infinitum*.'[3]

The scientific revolutionaries favoured the certainty of mathematics to the ambiguity of Scholastic accounts. Mathematics was 'built on clear and settled signification of names, which admit of no ambiguity'. This remark comes from Joseph Glanvill, whose defence of modern thought in *Scepsis Scientifica* earned him a place in the Royal Society in 1664. On Glanvill's account, Aristotle was exactly the opposite: 'Peripatetic philosophy is litigious'; its accounts are 'circular'; and its terms are 'empty', 'ambiguous', and lacking

[1] Though see Daniel Hausman on economics. Hausman 1992.
[2] Gassendi 1624 [1972].
[3] *Ibid.*, book 2, exercise 4, article 4.

'settled, constant signification'. The science of the Scholastics was involved in endless quarrels about words and very little actual investigation in large part because it tried to explain the behaviour of things by reference to their natures. But knowledge of natures, according to the new empiricists of the scientific revolution, is forever beyond our grasp; it is divine, not human. As Gassendi argues, it is not possible for mere humans to know 'that something is by nature and in itself, and as a result of basic, necessary infallible causes, constituted in a certain way'.[4] Rather, 'it can only be known how a thing appears to one or another'.[5]

It is on account of this twofold fact that the Aristotelians got into useless debates over meanings: on the one hand, natures stood at the core of explanation for them; on the other, these natures were intrinsically unknowable. According to the empiricists, then, the Aristotelians inevitably resolved things into qualities that were occult; these qualities could never be genuinely understood but only grasped by definition. Invariably this leads to circularity of explanation. The favoured example is that of gravity. Glanvill tells us:

> That heavy bodies descend by gravity is no better account than we would expect from a rustic; that gravity is a quality whereby a heavy body descends, is an impertinent circle, and teaches nothing.[6]

According to the empiricists we must throw over this attempt to found science on occult natures and instead base everything on the kinds of qualities that appear to us in experience. Even here there is a danger that we may become too ambitious, and Glanvill warns, 'If we follow manifest qualities beyond the empty signification of their names, we shall find them as occult as those which are professedly so.'[7]

Most modern accounts in the philosophy of science take it that the attempts of the scientific revolution to banish natures from science were successful. The idea of natures operating in things to determine their behaviours was replaced by the concept of a *law of nature*. Here is a short history, told by a modern-day empiricist, to illustrate:

> Aquinas was at pains to contest a preceding scholastic view that everything which happens does so because it is directly and individually willed by God. This would seem to make science a pointless enterprise; according to Aquinas it also denigrates creation. Yet theology points to God as ultimate cause. The reconciliation Aquinas offered was this: to explain why phenomena happen as they do, requires showing why they must; this necessity however derives from the natures of the individual substances involved which themselves are as they are because of God's decrees for

[4] *Ibid.*, book 6, article 1.
[5] *Ibid.*, book 6, article 6.
[6] Glanvill 1665 [1970], ch. 20.
[7] *Ibid.*

the world as a whole, made at the point of creation, but derives *proximately* from the local conditions and characters in the Aristotelian pattern

if we look more closely at the seventeenth century we see an insistence, even more adamant than Aquinas', upon the autonomy of physics from theology. Descartes insists on it most stringently

The *Drang nach Autonomie* of physics, even as developed by such theological thinkers as Descartes, Newton, and Leibniz, needed an intermediate link between God's decree and nature. Aquinas had needed such a link to explain proximate causation, and found it in the Aristotelian substantial forms (individual natures). For the seventeenth century another kind was needed, one that could impose a global constraint on the world process. In general terms, this link was provided by the idea that nature has its inner necessities, which are not mere facts, but constrain all mere facts into a unified whole. The theological analogy and dying metaphor of law provided the language in which the idea could be couched.[8]

My thesis here is that this story is distorted, at least as it applies to modern experimental science. We have not replaced *natures* by *laws of natures*. For our basic knowledge – knowledge of capacities – is typically about natures and what they produce. Instead what we have done is to replace occult powers by powers that are visible, though it may take a very fancy experiment to see them. This is already apparent in Francis Bacon. Bacon still employs the Aristotelian idea of natures or essences, but for him these are not hidden. Bacon looks for the explanatory essences, but he looks for them among qualities that are observable. Consider his hunt for the essence of heat.[9] He makes large tables of situations in which heat occurs, in which it is absent, and in which it varies by degrees. Instances agreeing in the Form of Heat include, for instance, rays of the sun; damp, hot weather; flames; horse dung; strong vinegar; and so forth. Then he looks to see what other quality is always present when heat is present and always absent when heat is lacking. In this way he finds the true, simple nature that constitutes heat: motion. The point is that Bacon still hopes to find the nature of heat, but among visible, not occult, qualities.

Modern explanation similarly relies on natures, I will argue, though modern natures are like Bacon's and unlike those of the Scholastics in that they are attributed to structures and qualities we can independently identify. Generally they differ from Bacon's in that they do not lie on the surface and are not to be observed with the naked eye. We often need very subtle and elaborate experiments in order to see them. Modern science insists that we found explanation on experimentally identifiable and verifiable structures and

[8] Van Fraassen 1989, pp. 4–6.
[9] Bacon 1620 [1994].

qualities. But, I maintain, what we learn about these structures and qualities is what it is in their natures to do.

What we have done in modern science, as I see it, is to break the connection between what the explanatory nature is – what it is, in and of itself – and what it does. An atom in an excited state, when agitated, emits photons and produces light. It is, I say, in the nature of an excited atom to produce light. Here the explanatory feature – an atom's being in the excited state – is a structural feature of the atom, which is defined and experimentally identified independently of the particular nature that is attributed to it. It is in the nature of the excited atom to emit light, but that is not what it is to *be* an atom in an excited state. For modern science what something really is – how it is defined and identified – and what it is in its nature to do are separate things. So even a perfect and complete modern theory would never have the complete deductive structure that the Aristotelians envisaged. Still, I maintain, the use of Aristotelian-style natures is central to the modern explanatory programme. We, like Aristotle, are looking for 'a cause and principle of change and stasis in the thing in which it primarily subsists', and we, too, assume that this principle will be 'in this thing of itself and not *per accidens*'.[10]

Yet even at this very cursory level of description we differ from Aristotle in three important ways. First, as in my example of an atom in an excited state, we assign natures not to substances but rather to collections or configurations of properties, or to structures. Second, like the early empiricists and the mechanical philosophers of the scientific revolution, modern physics supposes that the 'springs of motion' are hidden behind the phenomena and that what appears on the surface is a result of the complex interaction of natures. We no longer expect that the natures that are fundamental for physics will exhibit themselves directly in the regular or typical behaviour of observable phenomena. It takes the highly controlled environment of an experiment to reveal them. Third, having made the empiricist turn, we no longer identify natures with essences. As I have described in this section, in modern science we separate our definition of a property from our characterisation of what kind of change it naturally produces. Still, when we associate a particular principle of change with a given structure or characteristic, we expect that association to be permanent, to last so long as the structure is what it is. Indeed, it is this permanence of association that I underline by claiming that modern science still studies Aristotelian-style natures. Of course these are not really Aristotelian natures. For one thing, we seem to share none of the concerns about substance and individuation in which Aristotle's concept was embedded. There are a number of other differences as well. Nevertheless I

[10] Aristotle [1970], Book II.

call them *Aristotelian* because of the inheritance through the Scholastics to the 'New Philosophy' of Galileo, Bacon, and Descartes.

3 Natures and the analytic method

The modern empiricist account of regularities almost invariably presupposes a distinction crucial to the thought of Locke, Berkeley and Hume: the distinction between powers and sensible qualities. According to Hume, powers are not accessible to us through our senses and hence must be excluded from science. Nowadays the distinction takes a slightly different form, between the power things have to behave in certain ways, on the one hand, and the actually exhibited behaviours, on the other. But modern empiricists in the Hume tradition remain just as eager as Hume himself to reject powers. Laws of nature, they insist, are about *what things do*. I maintain, by contrast, that fundamental laws are generally not about what things do but what it is in their nature to do. Recall the discussion of Coulomb's law of electrostatic attraction and repulsion. Coulomb's law says that the force between two objects of charge q_1 and q_2 is equal to $q_1q_2/4\pi\varepsilon_0r^2$. Yet, this is not the force the bodies experience; they are also subject to the law of gravity. We say that Coulomb's law gives the force *due* to their charge. But this is no concept for an empiricist. What we mark when we say that there is a Coulomb force at work is not the presence of an occurrent force whose size is that given in Coulomb's law, but rather the fact that the charged bodies have *exercised their capacity* to repel or attract. Coulomb's is never the force that actually occurs.

I think the best account we can give is in terms of natures. Coulomb's law tells not what force charged particles experience but rather what it is in their nature, *qua* charged, to experience. Natures are closely related to powers and capacities. To say it is in their nature to experience a force of $q_1q_2/4\pi\varepsilon_0r^2$ is to say at least that they would experience this force if only the right conditions occur for the power to exercise itself 'on its own', for instance, if they have very small masses so that gravitational effects are negligible. But it is to say more: it is to say that their *tendency* to experience it persists even when the conditions are not right; for instance, when gravity becomes important. *Qua* massive, they tend to experience a different force, Gm_1m_2/r^2. What particles that are both massive and charged actually experience will depend in part upon what tendency they have *qua* charged and what *qua* massive.

It is to mark this fact, the fact that charge always 'contributes' the same force, that I use the Aristotelian notion of *nature*. But, as I remarked in referring to Bacon, these modern natures differ from Aristotle's in one very central respect. Although it is in the nature of charge to be subject to a force of $q_1q_2/4\pi\varepsilon_0r^2$, this nature does not in any proper Aristotelian way reveal the essence of charge. What charge is depends on a lot of factors independent of Coulomb's law. As Gerd Buchdahl puts it, there is a mere 'brute-fact connection' between what charge is and how charged particles behave *qua* charged.[11]

One customary response that Humeans make to the kinds of problems I am raising is to resort to counterfactuals. They talk not in terms of actually exhibited qualities and behaviours but instead of possible qualities and behaviours. Coulomb's law gives the force two bodies *would* experience if their masses were equal to zero. This seems a peculiar kind of counterfactual for an empiricist to place at the foundation of our study of motion, for it is one whose antecedent can never be instantiated. But that is not my principal concern here. Instead I want to point out another crucial nonempiricist element that is concealed in this account. Why do we want to know what the force between charged bodies would be *were the masses equal to zero*? This case is just one particular case among all conceivable ones and a peculiarly inconvenient one at that. Why, then, are these circumstances so special? They are special because these are the circumstances in which other hindrances are stripped away so that we can find out what charges do 'on their own' – that is, what charged particles do *by virtue of being charged*. This is how they would attract or repel one another were 'only' charge at work; and it is how they try to behave even when other factors impede them.

We discover the nature of electrostatic interaction between charges by looking in some very special circumstances. But the charge interaction carries that nature with it, from one circumstance to another. That is why what we call the *analytic method* works in physics: to understand what happens in the world, we take things apart into their fundamental pieces; to control a situation we reassemble the pieces, we reorder them so they will work together to make things happen as we will. You carry the pieces from place to place, assembling them together in new ways and new contexts. But you always assume that they will try to behave in new arrangements as they have tried to behave in others. They will, in each case, act in accordance with their nature.

The analytic method is closely associated with what we often call *Galilean idealisation*. Together idealisation and the inference to natures form a familiar two-tiered process that lies at the heart of modern scientific inquiry. First we try to find out by a combination of experimentation, calculation and inference how the feature under study behaves, or would behave, in a particular, highly specific situation. By controlling for or calculating away the gravitational effects, we try to find out how two charged bodies 'would interact if their masses were zero'. But this is just a stage; in itself this information is uninter-

[11] Buchdahl 1969.

esting. The ultimate aim is to find out how the charged bodies interact not when their masses are zero, nor under any other *specific* set of circumstances, but rather how they interact *qua* charged. That is the second stage of inquiry: we infer the nature of the charge interaction from how charges behave in these specially selected 'ideal' circumstances.

The key here is the concept *ideal*. On the one hand we use this term to mark the fact that the circumstances in question are not real or, at least, that they seldom obtain naturally but require a great deal of contrivance even to approximate. On the other, the 'ideal' circumstances are the 'right' ones – right for inferring what the nature of the behaviour is, in itself. Focusing on the first aspect alone downplays our problems. We tend to think that the chief difficulties come from the small departures from the ideal that will always be involved in any real experiment: however small we choose the masses in tests of Coulomb's law, we never totally eliminate the gravitational interaction between them; in Galilean experiments on inertia, the plane is never perfectly smooth nor the air resistance equal to zero; we may send our experiments deep into space, but the effect of the large massive bodies in the universe can never be entirely eliminated; and we can perform them at cryogenic temperatures, but the conditions will never, in fact, reach the ideal.

The problem I am concerned with is not whether we can get the system into ideal circumstances but rather, what makes certain circumstances ideal and others not. What is it that dictates which other effects are to be minimised, set equal to zero, or calculated away? This is the question, I maintain, that cannot be answered given the conventional empiricist account of scientific knowledge. If we consider any particular experiment, it may seem that the equipment we move about, the circumstances we contrive, and the properties we calculate away are ones that can be described without mentioning natures. But in each case, what makes that arrangement of equipment in those particular circumstances 'ideal' is the fact that these are the circumstances where the feature under study operates, as Galileo taught, without hindrance or impediment, so that its nature is revealed in its behaviour. Until we are prepared to talk in this way about natures and their operations, to fix some circumstances as felicitous for a nature to express itself and others as impediments, we will have no way of determining which principle is tested by which experiment. It is this argument that I develop in the next section.

Before turning to this argument I should make a point about terminology. My use of the terms *capacity* and *nature* are closely related. When we ascribe to a feature (like charge) a generic capacity (like the Coulomb capacity) by mentioning some canonical behaviour that systems with that capacity would display in ideal circumstances, then I say that that behaviour is *in the nature of* that feature. Most of my arguments about capacities could have been put in terms of natures had I recognised soon enough how similar capacities, as I see them, are to Aristotelian natures. On the other hand, the use of the term 'natures' would seem very odd in the contemporary philosophical literature on causation, and would probably divert attention from the central points I want to make there about capacities versus laws, so perhaps it is not such a bad idea to keep both terms.

4 How do we know what we are testing?

For anyone who believes that induction provides the primary building tool for empirical knowledge, the methods of modern experimental physics must seem unfathomable. Usually the inductive base for the principles under test is slim indeed, and in the best experimental designs, where we have sufficient control of the materials and our knowledge of the requisite background assumptions is secure, one single instance can be enough. The inference of course is never certain, nor irrevocable. Still we proceed with a high degree of confidence, and indeed, a degree of confidence that is unmatched in large-scale studies in the social sciences where we do set out from information about a very great number of instances. Clearly in these physics experiments we are prepared to assume that the situation before us is of a very special kind: it is a situation in which the behaviour that occurs is repeatable. Whatever happens in this situation can be generalised.

This particular kind of repeatability that we assume for physics experiments requires a kind of permanence of behaviour across varying external conditions that is comparable to that of an essence although not as strong. For example, we measure, successfully we think, the charge or mass of an electron in a given experiment. Now we think we know the charge or mass of all electrons; we need not go on measuring hundreds of thousands. In so doing we are making what looks to be a kind of essentialist assumption: the charge or mass of a fundamental particle is not a variable quantity but is characteristic of the particle so long as it continues to be the particle it is.

In most experiments we do not investigate just the basic properties of systems, such as charge, but rather more complicated trains of behaviour. Diagrammatically we may think of Galileo's attempts to study the motions of balls rolling down inclined planes; or, entirely at the opposite end of the historical spectrum, the attempts in Stanford's Gravity-Probe-B experiment to trace the precession of four gyroscopes in space in order to see how they are affected by the space-time curvature relativistically induced by the earth. Here too some very strong assumptions must back our willingness to draw a general conclusion from a very special case. On the surface it may seem that the licence to generalise in these cases can be put in very local terms that need no reference to natures. We require only the assumption that all systems so situated as the one in hand will behave identically. But on closer inspection we can see that this is not enough.

We may begin to see why by considering Hume himself. Hume maintained the principle 'same cause, same effect'. For him every occurrence is an exemplar of a general principle. It is simply a general fact about the world, albeit one we can have no sure warrant for, that identically situated systems behave identically. Hence for Hume the licence to generalise was universal. But not for us. We cannot so easily subscribe to the idea that the same cause will always be succeeded by the same effect. Hume assumed the principle to be true, though not provable. He worried that principles like this one could only be circularly founded, because they could have no evidence that is not inductive. But nowadays we question not just whether our belief in them can be well-founded but whether they are true.

Even if we were content with merely inductive warrant, in what direction does our evidence point? The planetary motions seem regular, as do the successions of the seasons; but in general nature in the mundane world seems obstinately unruly. Outside the supervision of a laboratory or the closed casement of a factory-made module, what happens in one instance is rarely a guide to what will happen in others. Situations that lend themselves to generalisations are special, and it is these special kinds of situations that we aim to create, both in our experiments and in our technology. My central thesis in this chapter is that what makes these situations special is that they are situations that permit a stable display of the nature of the process under study, or the stable display of the interaction of several different natures.

The case is especially strong when we turn from fictional considerations of ideal reasoning to considerations of actual methodology. Here questions of true identity of circumstance drop away. We never treat complete descriptions; rather we deal with *salient characteristics* and *relevant similarities*. This is a familiar point. You do not have to specify everything. If the right combination of factors is fixed, you are in a position to generalise. Yet what makes a specific combination a right one? What is the criterion that makes one similarity relevant and another irrelevant? Experiments are designed with intense care and precision. They take hard work and hard thought and enormous creative imagination. The Gravity-Probe experiment which I mentioned is an exaggerated example. It will only be set running twenty-five years – twenty-five years of fairly continuous effort – after it was initiated, and it will have involved teams from thirty or forty different locations, each solving some separate problem of design and implementation.

What can account for our effort to make the experimental apparatus *just so* and no other way? Take the Gravity Probe as a case in point. Each effort is directed to solving a specific problem. One of the very first in the Gravity

Probe involved choosing the material for the gyroscopes. In the end they are to be made of fused quartz, since fused quartz can be manufactured to be homogeneous to more than one part in 10^6. The homogeneity is crucial. Any differences in density will introduce additional precessions, which can be neither precisely controlled nor reliably calculated, and these would obscure the nature of the general-relativistic precession that the experiment aims to learn about.

In this case we can imagine that the physicists designing the experiment worked from the dictum, which can be formulated without explicit reference to natures, 'If you want to see the relativistic precession, you had better make the gyroscope as homogeneous as possible.' They wanted to do that because they wanted to eliminate other sources of precession. But more than that is necessary. The total design of the experiment must take account not only of what else might cause precession but also of what kinds of features would interfere with the relativistic precession, what kinds of factors could inhibit it, and what is necessary to ensure that it will, in the end, exhibit itself in some systematic way. When all these factors are properly treated, we should have an experiment that shows what the nature of relativistic precession is. That is the form, I maintain, that the ultimate conclusion will take.

But that is not the immediate point I want to make. What I want to urge is that, by designing the experiment to ensure that the nature of relativistic precession can manifest itself in some clear sign, by blocking any interference and by opening a clear route for the relativistic coupling to operate unimpeded – *to operate according to its nature*, by doing just this, the Gravity-Probe team will create an experiment from which it is possible to infer a general law. At the moment the form of this law is not my chief concern. Rather, what is at stake is the question, 'What must be true of the experiment if a general law of any form is to be inferred from it?' I claim that the experiment must succeed at revealing the nature of the process (or some stable consequence of the interaction of natures) and that the design of the experiment requires a robust sense of what will impede and what will facilitate this. The facts about an experiment that make that experiment generalisable are not facts that exist in a purely Humean world.

It is, of course, not really true that my thesis about the correct form of natural laws is irrelevant to my argument. Put in the most simple-minded terms, what I point out is the apparent fact that we can generalise from a single observation in an experimental context just because that context is one in which all the relevant sources of variation have been taken into account. Then, after all, what I claim is that it is laws in the form I commend – that is, laws about natures – that determine what is and what is not relevant. This sets the obvious strategy for the Humean reply: laws, in the sense of universal or probabilistic generalisations, determine the relevant factors an experiment must control to ensure it is repeatable. I will discuss this strategy briefly in the next section. Before turning to it, though, I want to make some clarifications about the concept of generalisability.

I have been using the term 'generalisable' and the term 'repeatable'. Both can be taken in two senses in this discussion. I claim that the Gravity Probe aims to establish a general claim about the nature of the coupling of a spinning gyroscope to curved space-time and thereby to learn something about the truth of the general theory of relativity. But along the way, as a by-product, the experiment will reveal, or instantiate, another considerably less abstract claim, a claim that can far more readily be cast into the conventional form of a universal generalisation. This is a generalisation to the effect that any fused-quartz gyroscope of just this kind – electromagnetically suspended, coated uniformly with a very, very thin layer of superfluid, read by a SQUID detector, housed in a cryogenic dewar, constructed just so . . . and spinning deep in space – will precess at the rate predicted. We expect a claim like this to obtain because we expect the experiment to establish a stable environment in which whatever happens would happen regularly; that is, we take the experimental design to be a design for a nomological machine.

This is a sense of repeatability internal to the experiment itself: given that the experiment is a good one, if it were to be rerun in the same way with the same apparatus, it should generate the same behaviour. We need not demand that the regularity instantiated be expressible in some particular language – or in any language, for that matter; nor, as Harry Collins stresses,[12] need we insist that the knowledge of how to build the apparatus be explicit knowledge that could be read from the experimenter's notebooks or that could be written in a 'how-to-build-it' manual.[13] Yet if the experiment is to bear on the more general conclusion which we in the end want to establish, we do want to insist on the regularity. For part of what is meant by the hypothesis that the coupling between the gyroscope and the curvature has a special nature that bears on the truth of general relativity is that there is a proper, predictable way in which it will behave, if only the circumstances are propitious. To the degree that we doubt that the experiment is repeatable, to that degree at least must we doubt that the behaviour we see is a sign of the nature we want to discover.

Although the general (albeit low-level) law that expresses this first kind of repeatability is, it seems, a universal generalisation of the conventional form, still the argument I want to make for the necessity of some nonstandard forms

[12] Collins 1985.

[13] Still, the knowledge cannot be too implicit. Trivially, where the experiment is to serve as a test, we must know enough to be assured that the behaviour we see is a manifestation of the nature of the phenomenon we want to study and not a manifestation of some other side aspect of the arrangement.

in the background bears on it just as forcefully as on the more abstract law that seems directly to describe the natures. As with the higher-level law, so too with the lower-level: if we want to understand why we are entitled to accept this law on such a thin inductive base as the Gravity-Probe's four gyroscopes, and if we want to understand the painstaking details of design the experimenters labour over in order to produce the conditions of the law, we will have to use the idea of a *nature*, or some related non-Humean notion.

I want to make a fairly strong claim here. In the order of generality the low-level regularity claims about what happens in just this kind of experimental set-up come first, and the more abstract claim about the general nature of the coupling comes second. We tend to think that the order of warrant is parallel: the low-level generalisations come first and are most secure; the more abstract law about the nature of the coupling derives what warrant it gets from the acceptance of the generalisation. I want to urge that there is an aspect of warranting for which this picture is upside down. It is just to the extent that we acknowledge that the experiment is well designed to find out the natures of the interaction described in the higher-level claim that we are entitled to accept the low-level regularity on the basis of the experimental results.[14] We get no regularities without a nomological machine to generate them; and our confidence that *this* experimental set-up constitutes a nomological machine rests on our recognition that it is just the right kind of design to elicit the nature of the interaction in a systematic way.

This is the central argument with which this section began. But it bears repeating now that the distinction between low-level laws, in the form of generalisations, and high-level claims has been drawn. Most situations do not give rise to regular behaviour. But we can make ones that do. To do so, we deploy facts about the stable natures of the processes we manipulate and about the circumstances that will allow these natures either to act unimpeded or to suffer only impediments that can have a stable and predictable effect. When we have such a situation, we are entitled to generalise from even a single case.[15]

Return now to the two senses of repeatability. The first sense is internal to the specific experiment and bears on the low-level generalisation that is instanced there. The second sense crosses experiments and bears on the higher-level principle that is established: the results of an experiment should be repeatable in the sense that the high-level principles inferred from a par-

[14] I do not mean to suggest that there can be no other basis for this generalisation. Sheer repetition will serve as well, and that is an important aspect of claims like those of Hacking 1983, that the stable phenomena that are created in the experimental setting have a life of their own and continue to persist across great shifts in their abstract interpretation.

[15] Our degree of confidence in the generalisation will be limited, of course, by how certain we are that our assessment of the situation is correct.

ticular experiment should be borne out in different experiments of different kinds. In *Nature's Capacities and their Measurement* this kind of repeatability played a central role in arguing for the claim that these higher-level laws describe *natures*. Low-level generalisation is not enough. It is too tied to the specific details of the particular experiment; a generalisation about what occurs there simply does not cover what occurs elsewhere.

We might think that the problem arises merely from the fact that the language of these low-level laws is not abstract enough. We should not be talking about what happens to a spherically homogeneous ball of fused quartz, coated with a superconductor and spinning, electromagnetically suspended in midair. Rather, we should talk about a gyroscope and how it precesses. Still the move to more abstract language will not permit us to retain the simple, unproblematic form of a universal generalisation. For we do not want to record what all gyroscopes facing a significant space-time curvature *do*. Rather, we want to record what part the curvature coupling contributes to how a gyroscope precesses, no matter what, in the end, various and differently situated gyroscopes do. As I described in section 3, that is the core of the analytic method. The point is that we want to learn something from the experiment that is transportable to new situations where quite different circumstances obtain. We do that not by constructing regularity claims using super-abstract concepts but rather by learning the nature of the pieces from which the new situations are built.

I will not dwell on this argument. More about it can be found in *Nature's Capacities*. The argument I have wanted to make here is different. In *Nature's Capacities* I argue that we need something like natures if we are to generalise in the second sense – to infer from the results of one experiment some kind of claim that can cover other situations as well. Here I want to urge that we need the notion of natures to generalise in the first sense as well – to infer from the results of the experiment some general law that describes what happens, what happens in just this experimental situation whenever the experiment is run again. Returning to the remarks at the beginning of this section I may put the point another way. How do we know which generalisation, in this low-level sense, the experiment is testing? Not every feature of it is necessary to ensure its repeatability. The answer requires the notion of *natures*: the features that are necessary are exactly those which, in this very specific concrete situation, allow the nature of the process under study to express itself in some readable way. No weaker account will do. Without the concept of *natures*, or something very like it, we have no way of knowing what it is we are testing.

5 An objection

I have been arguing that in order to understand what makes experiments special, what ensures that we can generalise from them, we must employ concepts repugnant to a Humean, concepts such as *nature, power, impediment, operation*. The most obvious response for a Humean to make would be that the job can be equally well done by using only 'occurrent' properties and their regular associations.

So consider this response. We want to figure out what factors are relevant – what factors need to be controlled in a given experiment if that experiment is to be replicable. Imagine that we could be told whatever we wanted to know about the lawful regularities that would obtain in our universe if only enough repetitions occurred and pretend, for the sake of argument, that these regularities hold, as Humeans believe, *simpliciter*, not each relative to the successful operation of some nomological machine. What could we do to determine whether a given factor in our experiment is relevant or not and needs to be controlled? I suppose the procedure envisaged by the Humean is, very roughly, this: ask for all those laws whose consequents describe the same kind of behaviour (for example, *precessing in a gyroscope*) as that of the law we wish to infer from our experiment; any factor that appears in the antecedents of one of these laws is a relevant factor – that is, a factor that must be controlled in any experiment to test the law at hand. But at which level of law are we to conduct our search?

At the lower level there are a very great number of laws indeed. Gyroscopes of all shapes and materials and forms can precess, or fail to precess, in an inconceivable number of different determinate ways in a plentitude of different circumstances. The conditions are too numerous. They give us too many factors to control. Our experiments would be undoable and the laws they entitle would be narrowed in scope beyond all recognition. But there is a deeper problem: how are these laws to be read? For the Humean they must be the source of information about not only what factors are to be controlled but in exactly what way. Yet they cannot tell us that, for how a factor operates, at this very concrete level, is far too context-dependent. I give examples of this kind of context dependence elsewhere[16] but I think the point is easy to see. To know exactly what to do with the superconducting coating in the Gravity Probe, one needs to know about the detailed construction of that particular experiment; and the laws one wants to look at are not more laws about precession but rather laws about superconductors. The point is not whether these further laws are Humean in form or not but rather, how is the Humean to know to look at them? What is the prescription that sorts from

[16] *Cf.* Cartwright 1988 and 1989.

among all factors that appear in all the universal generalisations supposedly true in the world which factors are to be fixed, and how, in this particular experiment?

Perhaps the answer comes one level up. Here I think is where we get the idea that there might be a relatively small number of fixed, probably articulable, factors that are relevant. We may think in terms of forces, how few in kind they are; or of long lists of causes and preventives. What is crucial is that, at the higher level, context seems irrelevant. Either it is or it is not the case that magnetic fields deflect charged particles or that, as quantum mechanics teaches, an inversion in a population of molecules can cause lasing. Perhaps we can even find a sufficiently abstract law so that the problem seems to evaporate. For example if we are thinking of an experiment where the effect we look for involves particle motions, we turn to the law $\mathbf{F} = m\mathbf{a}$, and that tells us that we must control all sources of forces. I, of course, as we have seen in chapters 1 and 2, do not think this enough. We must control all unwanted causes of motion, and not all of these can be represented properly as forces. For the moment though, for the sake of argument, I am not pressing this point but will rather accept the more usual fundamentalist view in order to focus on the other problems for the Humean account. In the gyroscope experiment the law of choice in this case would be:

$$\text{Precession: } d(n^r_s)/dt = \Gamma^r \times n_s/I\omega_s$$

This formula gives the drift rate ($d(n^r_s)/dt$) of a gyrospin vector as a function of the total torque (Γ^r) exerted on the gyro along with its moment of inertia (I) and its spin angular velocity (ω_s). From this we learn: control all sources of torque except that due to the relativistic coupling, as well as any sources of deviation in the angular velocity and in the moment of inertia.

The difficulty with this advice is that it does not justify the replicability we expect unless we join to it a commitment to natures, or something very much like them. To see why, imagine a single successful run of the experiment, successful in the sense that, first, we have indeed managed to set the total net torque, barring that due to relativistic coupling, equal to zero – or, as the Gravity Probe hopes to do, at least to an order of magnitude lower than that predicted for the relativistic effect; and second, it turns out that the observed precession is just that predicted. We seem to have succeeded in giving a purely Humean recipe for when to generalise, and this case fits. Roughly, we can generalise the quantitative relation we see between a designated input (here the relativistic coupling) and the precession actually observed in a given situation if that situation sets the remaining net torque equal to zero (or, more realistically, calculates it away), where the rationale for picking *net torque* $= 0$ as the relevant feature comes from the 'Humean association' recorded in the functional law that describes the size of precessions.

The problem is that this does not get us the detailed generalisation we expect at the lower level. The Gravity-Probe team has worked hard to set the total net torque extremely low by a large number of specific hard-won designs; and they are entitled to think that the results are replicable in *that* experimental design. What the Humean prescription entitles them to is weaker. It gives them the right to expect only that on any occasion when the net nonrelativistic torque is zero, the precession will be the value predicted from the general theory of relativity. But we expect the more concrete general claim to hold as well.

Consider the table of design requirements for the gyroscope experiment (figure 4.1). The table tells how controlled each foreseeable source of torque must be in order for the total extraneous precession to be an order of magnitude smaller than that predicted from relativistic coupling. Each such source – rotor homogeneity, rotor sphericity, housing sphericity, optimum preload, and so on – presents a special design problem; and for each, the experiment has a special solution. Using fused quartz to get maximum rotor homogeneity is, for example, the starting point for the solution of the first problem. What all this careful planning, honing, and calculation entitles us to is a far more concrete generalisation than the one above about (near) zero external torque. We are entitled to infer from a successful run that in any experiment of this very *specific* design, the observed precession should be that predicted by the general theory of relativity.[17]

The table of requirements highlights the analytic nature of this kind of experiment. What happens if something goes wrong with the rotor housing as it was originally planned and the fault cannot be repaired? With a lot of effort the Probe team will make a new design and slot it into the old general scheme, making appropriate changes. Because we are working in a domain where we trust analytic methods, a peculiar kind of sideways induction is warranted: from the successful run with the original design plus our confidence in the new rotor housing and its placement, we are entitled to infer a second, highly specific 'lower-level' generalisation to the effect that the precession in situations meeting the new design will be that predicted for relativistic coupling as well. Again, the new situation will indeed be one that falls under the 'Humean' generalisation involving zero torques. What is missing

[17] The inference is *ceteris paribus*, of course – 'so long as nothing goes wrong'. The 'zero torque' generalisation apparently has the advantage that it needs no such *ceteris paribus* clause. But that is a mixed blessing since the advantage is bought at the cost of making 'zero torque' a concept that is not identifiable independently of its effect. As soon as we begin to fill in what *makes for* zero torque, anything we say will inevitably have to contain a *ceteris paribus* proviso, as I have argued in chapter 3.

			Requirement	Explanation
Gyroscope	Mechanical parts	Rotor homogeneity	$\frac{\Delta\rho}{\rho} \sim 3 \times 10^{-7}$	Mass unbalance and gravity gradient torques
		Rotor sphericity at rest	$\frac{\Delta r'}{r} \sim 5 \times 10^{-7}$ $\Delta r' \sim 0.4\,\mu in$	Suspension torques
		Housing sphericity	$\frac{\Delta r''}{r} \sim 2 \times 10^{-5}$ $\Delta r'' \sim 15\,in$	Second-order suspension torques and torques due to static charge on rotor
	Suspension systems	Optimum preload	$h \sim 10^{-7}$ g voltage $= 0.2$ V	Suspension torques and cyclic acceleration on rotor
		Preload symmetry	$\zeta \sim 1\%$	Suspension torques
		Centring accuracy	$t/d \sim 1\%$ $t = 15\,\mu in$	Suspension torques plus readout errors
	Other	Optimum spin speed	$\omega_s \sim 170$ Hz	Centrifugal distortion of rotor
		Torque switching ratio for spin up system	$\Gamma_r/\Gamma_s \sim 2 \times 10^{-13}$	See Section C(2)
		Distance from drag-free proof mass	$\ell < 20$ cm	To minimise cyclic accelerations on rotors

		Requirement	Explanation
Environment	Drag-free acceleration level	$<10^{-10}$ g	Suspension and mass unbalance torques
	Orbit eccentricity	<0.1	Suspension and mass unbalance torques
	Magnetic fields	$<10^{-7}$ g	Readout linearity (see Section C(3) (j) plus magnetic torques)
	Electric charge on ball	10^9 electrons 0.3 V	Torques plus preload requirement
	Gas presssure	Magnitude $< 10^{-9}$ torr Gradients $\sim 6\%$ variations allowed	Gas torques

Figure 4.1 Design requirements for a relativity gyroscope with limiting accuracy of 0.5×10^{-16} rad/sec (0.3 milliarc-sec/year). Source: Everitt 1980.

is the connection. The new situation is one of very small extraneous torque; but the expectation that it should be cannot be read from the regularities of nature.

The regularity theorist is thus faced with a dilemma. In low-level, highly concrete generalisations, the factors are too intertwined to teach us what will and what will not be relevant in a new design. That job is properly done in physics using far more abstract characterisations. The trouble is that once we have climbed up into this abstract level of law, we have no device within a pure regularity account to climb back down again.

6 An historical illustration: Goethe and Newton

So far I have couched the discussion in terms of making inductions from paltry samples, and that is because induction is the method that Humeans should favour for confirming laws. I think, though, that the process is far better understood as one of deduction. We accept laws on apparently slim experimental bases exactly when we can take for granted such strong background assumptions that (given these assumptions) the data plus the description of the experimental set-up deductively imply the law to be established. Probably the most prominent advocate of a deductive method in reasoning from experiment to law is Isaac Newton. It will be helpful to look briefly at Newton's use of the 'crucial experiment' in his theory of light and colours, and more particularly at Goethe's criticisms of it.

Newton's *experimentum crucis* is described in his first letter in 1671 to the Royal Society[18] in which he introduces his theory that white light consists of diverse rays of different refrangibility (that is, they are bent by different amounts when the light passes through a prism) and that colour is a property of the ray which depends on its refrangibility. The work reported in the letter is often taken as a model of scientific reasoning. Thomas Kuhn, for instance, claims that 'Newton's experimental documentation of his theory is a classic in its simplicity.'[19] According to Kuhn, the opposition view might eventually have accounted for some of the data that appeared to refute it, 'but how could they have evaded the implications of the *experimentum crucis*? An innovator in the sciences has never stood on surer ground.'[20]

It is important to keep in mind that Newton believed that his claims were *proven* by his experiments. In his letter he maintains, 'The Theory, which I propounded, was evinced by me, not inferring 'tis thus because not otherwise, that is, not by deducing it from a confutation of contrary suppositions but by

[18] Newton's first letter (1671) to the Royal Society, quoted from Newton 1959–76.

[19] Kuhn 1958, p. 36.

[20] *Ibid.*

deriving it from experiments concluding positively and directly.' Or, 'If the Experiments, which I urge, be defective, it cannot be difficult to show the defects; but if valid then by proving the theory they must render all objections invalid.' One last remark to illustrate the steadfastness of Newton's views on the role of the *experimentum crucis* in proving this claim appears in Newton's letter of 1676,[21] four years after his initial report to the Royal Society. This letter concerned the difficulties Anthony Lucas had reported in trying to duplicate Newton's experiments and also some of Lucas' own results that contradicted Newton's claims. Newton replies, 'Yet it will conduce to his more speedy and full satisfaction if he a little change the method he has propounded, and instead of a multitude of things try only the *Experimentum Crucis*. For it is not number of experiments, but weight to be regarded; and where one will do, what need many?'

Goethe's point of view is entirely opposite to Newton's: 'As worthwhile as each individual experiment may be, it receives its real value only when united or combined with other experiments . . . I would venture to say that we cannot prove anything by one experiment or even several experiments together.'[22] For Goethe, all phenomena are connected together, and it is essential to follow through from each experiment to another that 'lies next to it or derives directly from it'. According to Goethe, 'To follow every single experiment through its variations is the real task of the scientific researcher.' This is illustrated in his own work in optics where he produces long series of 'contiguous' experiments, each of which is suggested by the one before it. The point is not to find some single set of circumstances that are special but rather to lay out all the variations in the phenomena as the circumstances change in a systematic way. Then one must come to see all the interrelated experiments together and understand them as a whole, 'a single piece of experimental evidence explored in its manifold variations'.

Goethe is sharp in his criticisms of Newton. Two different kinds of criticism are most relevant here. The first is that Newton's theory fails to account for all the phenomena it should and that is no surprise since Newton failed to look at the phenomena under a sufficient range of variation of circumstance. Second, Newton's inferences from the experiments he did make were not valid; the *experimentum crucis* is a case in point. The chief fault which Goethe finds with Newton's inferences is one that could not arise in Goethe's method. Newton selects a single revealing experiment to theorise from; since he does not see how the phenomena change through Goethe's long sequences of experiments, he does not recognise how variation in circumstance affects the outcome: '[Newton's] chief error consisted in too quickly and hastily

[21] Newton's second letter (1676) to the Royal Society, quoted from Newton 1959–76.
[22] Goethe 1812 [1988].

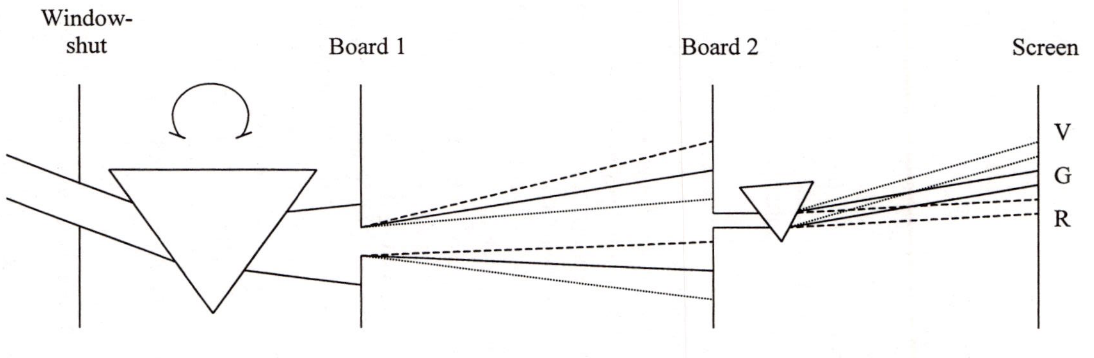

Figure 4.2 Newton's *experimentum crucis*. Source: Sepper 1988; recreated by G. Zouros.

setting aside and denying those questions that chiefly relate to whether external conditions cooperate in the appearance of colour, without looking more exactly into the proximate circumstances.'[23]

The crucial experiment involves refracting a beam of light through a prism, which elongates the initial narrow beam and 'breaks' it into a coloured band, violet at the top, red at the bottom. Then differently coloured portions of the elongated beam are refracted through the second prism. Consider figure 4.2, which is taken from Dennis L. Sepper's study, *Goethe contra Newton*. In all cases the colour is preserved, but at one end of the elongated beam the second refracted beam is elongated more than it is at the other. In each case there is no difference in the way in which the light falls on the prism for the second refraction. Newton immediately concludes, 'And so the true cause of the length of the image was detected to be no other than that light consists of *rays differently refrangible*.'[24]

We should think about this inference in the context of my earlier cursory description of the modern version of the deductive method, called 'bootstrapping' by Clark Glymour,[25] who has been its champion in recent debates. In the bootstrapping account, we infer from an experimental outcome to a scientific law, as Newton does, but only against a backdrop of rather strong assumptions. Some of these assumptions will be factual ones about the specific arrangements made – for example, that the angle of the prism was 63°; some will be more general claims about how the experimental apparatus works – the theory of condensation in a cloud chamber, for instance; some will be more general claims still – for example, all motions are produced by forces; and some will be metaphysical, such as the 'same cause, same effect' principle mentioned above. The same is true of Newton's inference. It may be a perfectly valid inference, but there are repressed premises. It is the repressed premises that Goethe does not like. On Goethe's view of nature, they are not only badly supported by the evidence; they are false. Colours, like all else in Goethe's world,[26] are a consequence of the action of opposites, in this case light and darkness:

We see on the one side light, the bright; on the other darkness, the dark; we bring

[23] Sepper 1988, p. 144.

[24] Newton's first letter (1671) to the Royal Society, quoted from Newton 1959–76.

[25] Glymour 1980.

[26] See, for example, paragraph 739 of the didactic part of his *Theory of Colour*, 1810 [1967]: 'The observers of nature, however they may differ in opinion in other respects, will agree that all which presents itself as appearance, all that we meet with as phenomenon, must either indicate an original division which is capable of union, or an original unity which admits of division, and that the phenomenon will present itself accordingly. To divide the united, to unite the divided, is the life of nature; this is the eternal systole and diastole, the eternal collapsion and expansion, the inspiration and expiration of the world in which we live and move.'

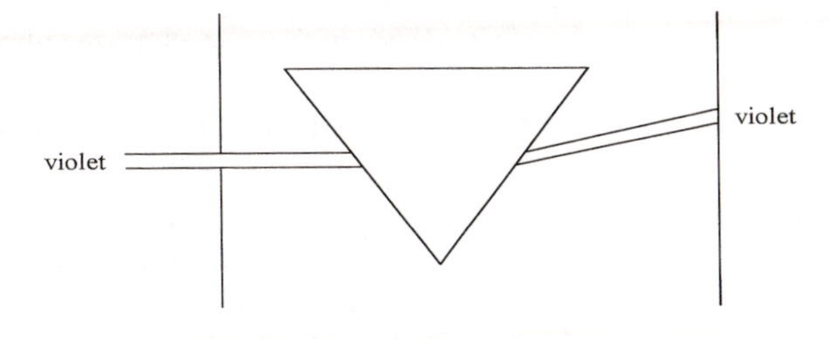

Figure 4.3.

what is turbid between the two [such as a prism or a semitransparent sheet of paper], and out of these opposites, with the help of this mediation, there develop, likewise in an opposition, colors.[27]

Newton's argument requires, by contrast, the assumption that the tendency to produce colours is entirely in the nature of the light, and that is why this dispute is of relevance to my point here. As Sepper says, for Newton 'the cause is to be sought only in the light itself'.

Let us turn to Newton's reasoning. The argument is plausible, so long as one is not looking for deductive certainty. From Newton's point of view (though not from that of Goethe, who imagines a far richer set of possibilities), the two hypotheses to be decided between are: (a) something that happens involving white light in the prism produces coloured light; or (b) coloured light is already entering the prism in the first place. We can see the force of the argument by thinking in terms of inputs and outputs. Look at what happens to, say, the violet light in the second prism (figure 4.3) and compare this with the production of violet light in the first prism (figure 4.4). In both cases the outputs are the same. The simplest account seems to be that

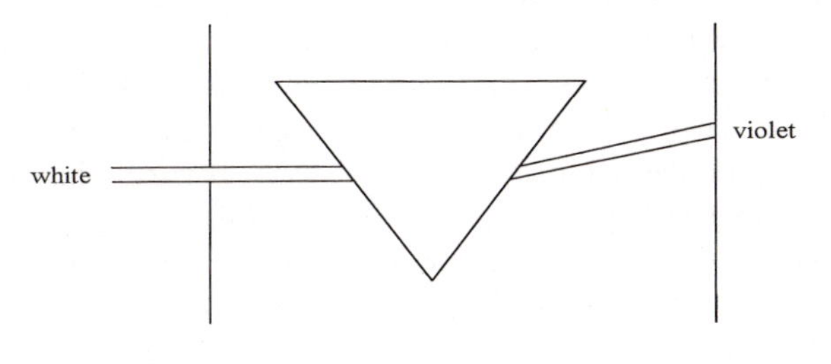

Figure 4.4.

[27] Goethe 1810 [1967], didactic part, paragraph 175.

the prism functions in the same way in both cases: it just transports the coloured light through, bending it in accord with its fixed degree of refrangibility.

Consider an analogous case. You observe a large, low building. Coloured cars drive through. Cars of different colours have different fixed turning radii. You observe for each colour that there is a fixed and colour-dependent angle between the trajectory on which the car enters the building and the trajectory on which it exits; moreover, this is just the angle to be expected if the cars were driven through the building with steering wheels locked to the far left. Besides cars, other vehicles enter the building, covered; and each time a covered vehicle enters, a coloured car exits shortly afterward. It exists at just that angle that would be appropriate had the original incoming vehicle been a car of the same colour driven through with its steering wheel locked. Two hypotheses are offered about what goes on inside the building. Both hypotheses treat the incoming coloured cars in the same way: on entering the building their steering wheels get locked and then they are driven through. The two hypotheses differ, however, about the covered vehicles. The first hypothesis assumes that these, too, are coloured cars. Inside the building they get unwrapped, and then they are treated just like all the other coloured cars. The second hypothesis is more ambitious. It envisages that the low building contains an entire car factory. The covered vehicles contain raw material, and inside the building there are not only people who lock steering wheels, but a whole crew of Fiat workers and machinery turning raw materials into cars.

Obviously, the first hypothesis is simpler, but it has more in its favour than that. For so far, the second hypothesis has not explained why the manufactured cars exit at the angle they do, relative to their incoming raw materials; and there seems to be no immediate natural account to give on the second story. True, the cars are manufactured with fixed turning radii, but why should they leave the factory at just the same angle relative to the cart that carries in their raw materials as a drive-through does relative to its line of entry? After all, the manufactured car has come to exist only somewhere within the factory, and even if its steering wheel is locked, it seems a peculiar coincidence should that result in just the right exit point to yield the required angle *vis-à-vis* the raw materials. In this case, barring other information, the first, Newtonian, hypothesis seems the superior. The caveat, 'barring other information', is central, of course, to Goethe's attack. For, as I have already remarked, Goethe was appalled at the small amount of information that Newton collected, and he argued that Newton's claim was in no way adequate to cover the totality of the phenomena. What looks to be the best hypothesis in a single case can certainly look very different when a whole array of different cases have to be considered.

The principal point to notice, for my purpose, is that the argument is not at all deductive. It can only become so if we already presuppose that we are looking for some fixed feature in light itself that will account for what comes out of the prism – something, as I would say, in the nature of light. Any assumption like this is deeply contrary to Goethe's point of view. The first few paragraphs of Newton's letter, before the introduction of the crucial experiment, give some grounds for such an assumption on his part; Goethe makes fun of them:

> It is a fact that under *those circumstances* that Newton exactly specifies, the image of the sun is five times as long as it is wide, and that this elongated image appears entirely in colors. Every observer can repeatedly witness this phenomenon without any great effort.
>
> Newton himself tells us how he wants to work in order to convince himself that no *external cause* can bring this elongation and coloration of the image. This treatment of his will, as already was mentioned above, be subjected to criticism for we can raise many questions and investigate with exactness, whether he went to work properly and to what extent his proof is in every sense complete. If one analyzes his reasons, they have the following form: When the ray is refracted the image is longer than it should be according to the laws of refraction.
>
> Now I have tried everything and thereby convinced myself that no external cause is responsible for this elongation.
>
> Therefore it is an inner cause, and this we find in the divisibility of light. For since it takes up a larger space than before, it must divided, thrown asunder; and since we see the sundered light in colors, the different parts of it must be colored.
>
> How much there is to object to immediately in this rationale![28]

The contrast that I want to highlight is between Newton's postulation of an inner cause in light versus Goethe's long and many-faceted row of experiments. Goethe often remarks that he and Newton both claim to be concerned with *colors*; Newton after all labels his account in the 1671 letter his 'new theory of light and colors'. But, in actuality, Goethe points out, Newton's work is almost entirely about the behaviour of rays – that is, about the inner nature of light. Goethe's experiments often involve light, but it is not light that he studies. The experiments describe entire interacting complexes, such as evening light entering a room through a hole in a white blind on which a candle throws light ('snow seen through the opening will then appear blue, because the paper is tinged with warm yellow by the candlelight'[29]), or sunlight shining into a diving bell (in this case 'everything is seen in a red light . . . while the shadows appear green'[30]), or a particularly exemplary case for the existence of coloured shadows, a pencil placed on a sheet of white paper between a short, lighted candle and a window so that the twilight from the

[28] Goethe 1793; quoted from Sepper 1988, p. 101.
[29] Goethe 1810, didactic part, paragraph 79.
[30] *Ibid.*, didactic part, paragraph 78.

window illuminates the pencil's shadow from the candle ('the shadow will appear of the most beautiful blue'[31]). Even when described from the point of view of Goethe's final account of colour formation, in the prism experiments Goethe is not looking at light but rather at light (or darkness)-in-interaction-with-a-turbid-medium.

Newton focuses on his one special experiment and maintains that the account of the phenomena in that experiment will pinpoint an explanation that is generalisable. The feature that explains the phenomena in that situation will explain phenomena in other situations; hence he looks to a feature that is part of the inner constitution of light itself. To place it in the *inner* constitution is to cast it not as an observable property characteristic of light but rather as a power that reveals itself, if at all, in appropriately structured circumstances. To describe it as part of light's *constitution* is to ascribe a kind of permanence to the association: light retains this power across a wide variation in circumstance – indeed, probably so long as it remains light. That is, I maintain, to treat it as an Aristotelian-style nature. This is why Newton, unlike Goethe, can downplay the experimental context. The context is there to elicit the nature of light; it is not an essential ingredient in the ultimate structure of the phenomenon.

7 Who has dispensed with natures?

My argument in this chapter hinges on a not surprising connection between methodology and ontology. If you want to find out how a scientific discipline pictures the world, you can study its laws, its theories, its models, and its claims – you can listen to what it says about the world. But you can also consider not just what is said but what is done. How we choose to look at the world is just as sure a clue to what we think the world is like as what we say about it. Modern experimental physics looks at the world under precisely controlled or highly contrived circumstance; and in the best of cases, one look is enough. That, I claim, is just how one looks for natures and not how one looks for information about what things do.

Goethe criticises Newton for this same kind of procedure that we use nowadays, and the dispute between them illustrates my point. Newton's conclusions in his letter of 1671, as well as throughout his later work in optics, are about the inner constitution of light. I claim that this study of the inner constitution is a study of an Aristotelian-style nature and that Newton's use of experiment is suited to just that kind of enterprise, where the *experimentum crucis* is an especially striking case. The coloured rays, with their different degrees of refrangibility, cannot be immediately seen in white light. But

[31] *Ibid.*, didactic part, paragraph 65.

through the experiment with the two prisms, the underlying nature expresses itself in a clearly visible behaviour: the colours are there to be seen, and the purely dispositional property, *degree-of-refrangibility*, is manifested in the actual angle through which the light is bent. The experiment is brilliantly constructed: the connection between the natures and the behaviour that is supposed to reveal them is so tight that Newton takes it to be deductive.

Goethe derides Newton for surveying so little evidence, and his worries are not merely questions of experimental design: perhaps Newton miscalculated, or mistakenly assumed that the second prism was identical in structure with the first, or Newton takes as simple what is not . . . Goethe's disagreement with Newton is not a matter of mere epistemological uncertainty. It is rather a reflection of deep ontological differences. For Goethe, all phenomena are the consequence of interaction between polar opposites. There is nothing in light to be isolated, no inner nature to be revealed. No experiment can show with a single result what it is in the nature of light to do. The empiricists of the scientific revolution wanted to oust Aristotle entirely from the new learning. I have argued that they did no such thing. Goethe, by contrast, did dispense with natures; there are none in his world picture. But there are, I maintain, in ours.

ACKNOWLEDGEMENTS

This chapter is a slightly shortened version of Cartwright 1992. I owe special thanks to Hasok Chang, as well as to the members of the Philosophy of Language and Science Seminar and the Romanticism and Science Seminar at Stanford University, fall and winter terms, 1989/90. This chapter is dedicated to Eckart Förster.

5 Causal diversity; causal stability

1 Particularism and the hunt for causal laws

This book takes its title from a poem by Gerard Manley Hopkins. Hopkins was a follower of Duns Scotus; so too am I. I stress the particular over the universal and what is plotted and pieced over what lies in one gigantic plane. This book explores the importance of what is different and separate about the sciences as opposed to what is common among them. The immediate topic of the present chapter is the *Markov condition*. This condition lies at the heart of a set of very well grounded and powerful techniques for causal inference developed during the last fifteen years by groups working at Carnegie Melon University with Clark Glymour and Peter Spirtes[1] and at UCLA with Judea Pearl[2] using directed acyclic graphs (DAGs). I think of it as a small case study in particularism. Contrary to what I take to be the hopes of many who advocate DAG-techniques, I argue that the methods are not universally applicable, and even where they work best, they are never sufficient by themselves for causal inferences that can ground policy recommendations. For sound policy, we need to know not only what causal relations hold but what will happen to them when we undertake changes. And that requires that we know something about the design of the nomological machine that generates these causal relations.

Section 2 of this chapter points out that the Markov condition is in general not satisfied by causes that act probabilistically, though it may be appropriate for many kinds of deterministic models. I use this to illustrate that there is a great variety of different kinds of causes and that even causes of the same kind can operate in different ways. Consider a causal law of the form '*X* causes *Y*': *e.g.*, 'Sparking an inverted population causes coherent radiation.' This could be true because sparking precipitates the coherent radiation, which is already waiting in the wings; or because it removes a prior impediment; or because it acts to create a coherent radiation that would never have been

[1] *Cf.* Spirtes et al. 1993.
[2] *Cf.* Pearl 1993, 1995 and forthcoming.

possible otherwise. Does the sparking cause coherent radiation in a way that also tends to decohere it? Does it do so deterministically? Or probabilistically? The term 'cause' is highly unspecific. It commits us to nothing about the kind of causality involved nor about how the causes work. Recognising this should make us more cautious about investing in the quest for universal methods for causal inference. *Prima facie*, it is more reasonable to expect different kinds of causes operating in different ways or embedded in differently structured environments to be tested by different kinds of statistical tests.

Section 3 rehearses some problems generated for the Markov condition by *mixing*, that is by combining together different populations in which either different causal relations hold for the same set of factors or different probability measures. These problems are well known; the fact that there can be different causal relations among the very same features in different circumstances as well as different probabilities is widely recognised. This suggests that neither the causal relations studied by these techniques nor the probabilities are fundamental. Rather, as I have been arguing, both depend on the stable operation of a nomological machine. The most straightforward way to try to save the Markov condition in the face of the mixing problem is to treat the design of the machine itself as just another cause, adding to the Book of Nature a great number of new law claims in which it is supposed to figure, both causal laws and probabilities, and always treating the new law claims as just the same in type as the old ones. Reflecting on the kinds of peculiarities and difficulties that this strategy creates will help us to understand why we need new concepts and new methods to understand where the laws of nature come from.

2 No models in; no causes out

2.1 The Markov condition

Although the techniques I discuss here have been developed both by Pearl's group and by the Glymour/Spirtes group, I will centre my discussion on the latter, which I know better, and in particular on the 1993 book, *Causation, Prediction and Search* by Peter Spirtes, Clark Glymour and Richard Scheines. Spirtes, Glymour and Scheines assume that causality and probability are distinct but related features of the world. They suppose that the causal laws relating a set of variables can be represented in a directed acyclic graph like that in figure 5.1, and in addition that Nature provides a probability measure over these variables. The first job is to say what relations can be assumed to hold between the causal structure depicted in a directed graph and the associated probability measure. Given these relations, Spirtes,

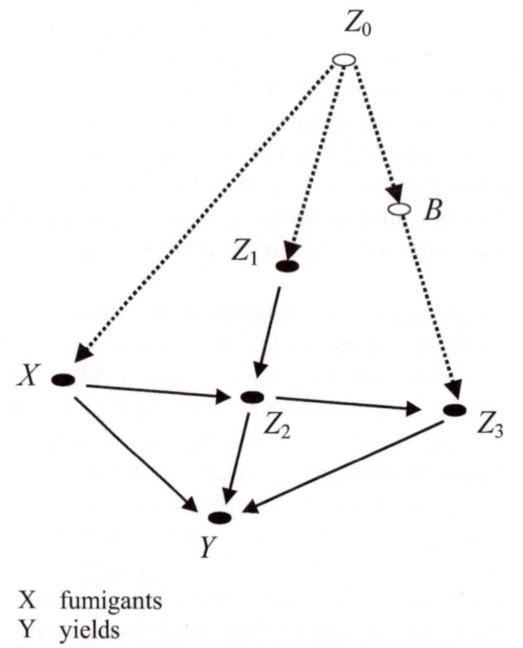

X fumigants
Y yields
B the population of birds and other predators
Z_0 last year's eelworm population
Z_1 eelworm population before treatment
Z_2 eelworm population after treatment
Z_3 eelworm population at the end of the season

Figure 5.1 A causal diagram representing the effect of fumigants, X, on yields, Y. Source: Pearl 1995, p. 670.

Glymour and Scheines generate mechanical procedures for filling in arrows in the structure given information about the probabilities. And they have a very nice theorem that says that given a causal structure and a probability measure that match in the right way, these procedures will never fill in a causal arrow on the graph that is not there in the real world. The motivation, I take it, is methodological. Probabilities are sometimes easier to find out about than are causes (at least for the statistician!), so we would like to understand how to use statistics to discover causes. The Markov condition is one of the central assumptions that constrain how probabilities and causes relate.

The Markov condition invokes the notion of a causal parent. In graphs like that in figure 5.1, the causal parents of a variable are all those at the open end of arrows heading into the variable. Clearly what is and what is not a causal parent is dependent on the choice of representation and not just on what causal laws hold. To represent *X* as a causal parent of *Y* is to claim that '*X* causes *Y*'. But not all causes will be represented and which causes appear as parents depends in part on the temporal graining chosen (though sometimes, as I argue later, it may not be possible to refine the graining more closely). The Markov condition says that, once we condition on all its parents, a variable will be probabilistically independent of all other variables except its descendants. It thus combines in one a prohibition against causes exerting influences across temporal gaps with the demand that parents of a variable render it independent from anything that follows after it except its own descendants. The first is what one might more naturally call a 'Markov' condition. The second is what philosophers call *screening off*, of which Hans Reichenbach's condition of the common cause is a special case: a full set of parents screens off the joint effects of any of these parents from each other. It is this latter aspect of the Markov condition that I shall be concerned with in this chapter.

2.2 Probabilistic versus deterministic causes

Because of the central role that probabilities play, one might think that the methods employed by Pearl's group and by Spirtes, Glymour and Scheines are methods that can allow us to study probabilistic causes, causes that operate purely probabilistically, without determinism in the offing. This is not entirely so. Why, after all, should we accept the Markov condition? That is the question that starts off section 3.5 of *Causation, Prediction and Search*: 'When and why should it be thought that probability and causality together satisfy these conditions, and when can we expect the conditions to be violated?'[3] The next sentence but one gives one of their two principal answers. 'If we consider probability distributions for the vertices of causal graphs of deterministic or pseudo-indeterministic systems in which the exogenous variables are independently distributed, then the Markov condition must be satisfied.'[4] (The exogenous variables in a graph are the ones with no causal inputs pictured in the graph.)

To understand what this means we must realise that what we can hope for with one of these graphs is accuracy; we never expect completeness. So we want a graph whose causal claims are a correct though incomplete representation of the causal structure that obtains in the world. Consider the relation of a particular effect to its causal parents. On most graphs this relation will be indeterministic: the values of the parent variables do not fix the value of the

[3] Spirtes et al. 1993, p. 57.
[4] *Ibid.*

effect variable. But the indeterminism may not be genuine. The structure will be *pseudo-indeterministic* if it can be embedded in another more complete graph that also makes only correct claims and in which the parents of the given effect are sufficient to fix the value of the effect.

When we do have a graph in which the parents of a given effect fix its value in this way, then the Markov condition is trivially true, since the probabilities are always either 0 or 1. But deterministic causes of this kind seem to be few and far between in nature, and it is important to be clear that the Markov condition will not in general hold for genuine probabilistic causality, where a richer set of choices confronts Nature. When causality is probabilistic, a question arises for Nature that does not have to be addressed for purely deterministic causes: what are the probabilistic relations among the operations of the cause to produce its various effects? In the case of determinism the answer is trivial. Whenever the cause occurs, it operates to produce each and every one of its effects. But when it can occur and nevertheless not operate, some decision is required about how to marshal its various operations together.

Screening off offers one special solution – the one in which all operations occur independently of each other. Sociology regularly provides us with cases modelled like this. Consider the familiar example of Hubert Blalock's negative correlation between candy consumption and divorce, which disappears if we condition on the common causal parent, age. Age operates totally independently in its action to decrease candy consumption from its operation to increase the chance of divorce. This is a kind of 'split-brain' model of the common cause. The single cause, age, sets about its operations as if it were two separate causes that have nothing to do with each other. It is as if the cause had two independent agents in one head. But the split-brain is by no means the most typical case. We confront all different kinds of joint operations everyday, both in ordinary life and, more formally, throughout the social sciences.

Consider a simple example. Two factories compete to produce a certain chemical that is consumed immediately in a nearby sewage plant. The city is doing a study to decide which to use. Some days chemicals are bought from Clean/Green; others from Cheap-but-Dirty. Cheap-but-Dirty employs a genuinely probabilistic process to produce the chemical. The probability of getting the desired chemical on any day the factory operates is eighty per cent. So in about one-fifth of the cases where the chemical is bought from Cheap-but-Dirty, the sewage does not get treated. But the method is so cheap the city is prepared to put up with that. Still they do not want to buy from Cheap-but-Dirty because they object to the pollutants that are emitted as a by-product whenever the chemical is produced.

That is what is really going on. But Cheap-but-Dirty will not admit to it.

They suggest that it must be the use of the chemical in the sewage plant itself that produces the pollution. Their argument relies on the screening-off condition. If the factory *were* a common parent, C, producing both the chemical X and the pollutant Y, then (assuming all other causes of X and Y have already been taken into account) conditioning on which factory was employed should make the chemical probabilistically independent from the pollutant, they say. Their claim is based on the screening-off condition. Screening off requires that conditioning on the joint case C should make the probabilities for its two effects factor. We should expect then:

$$\text{Prob}(XY/C) = \text{Prob}(X/C)\cdot\text{Prob}(Y/C)$$

Cheap-but-Dirty is indeed a cause of the chemical X, but they cannot then be a cause of the pollutant Y as well, they maintain, since

$$\text{Prob}(XY/C) = 0.8 \neq (0.8) \times (0.8) = \text{Prob}(X/C)\cdot\text{Prob}(Y/C)$$

Indeed the conditional probabilities do not factor, but that does not establish Cheap-but-Dirty's innocence. In this case it is very clear why the probabilities do not factor. Cheap-but-Dirty's process is a probabilistic one. Knowing that the cause occurred will not tell us whether the product resulted or not. Information about the presence of the by-product will be relevant since this information will tell us (in part) whether, on a given occasion, the cause actually 'fired'.

We may think about the matter more abstractly. Consider a simple case of a cause, C, with two separate yes-no effects, X and Y. Looking at the effects, we have in this case an event space with four different outcomes:

$$+X+Y,\ -X+Y,\ +X-Y,\ -X-Y$$

If causality is to be fully probabilistic, Nature must set probabilities for each of these possible outcomes. That is, supposing C occurs, Nature must fix a joint probability over the whole event space:

$$\text{Prob}_C(+X+Y),\ \text{Prob}_C(-X+Y),\ \text{Prob}_C(+X-Y),\ \text{Prob}_C(-X-Y)$$

Nothing in the concept of causality, or of probabilistic causality, constrains how Nature must proceed. In the case of two yes-no variables, the screening-off condition is satisfied only for a very special case, the one in which

$$\text{Prob}_C(+X+Y)\cdot\text{Prob}_C(-X-Y) = \text{Prob}_C(+X-Y)\cdot\text{Prob}_C(-X+Y)$$

Lesson: where causes act probabilistically, screening off is not valid. Accordingly, methods for causal inferences that rely on screening off must be applied with judgement and cannot be relied on universally.

I have often been told in response by defenders of the Markov condition: 'But the macroscopic world to which these methods are applied *is* deterministic. Genuine probabilistic causality enters only in the quantum realm.' Spirtes, Glymour and Scheines do not say this anywhere in the book. Still, the fact that the Markov condition is true of pseudo-indeterministic systems is one of their two chief reasons for thinking the Markov condition holds for 'macroscopic natural and social systems for which we wish causal explanations'.[5] This argument supposes, I think wrongly, that there are no macroscopic situations that can be modelled well by quantum theory but not by classical theory. But that is a minor point. Far more importantly, it confuses the world with our theories of it. Classical mechanics and classical electromagnetic theory are deterministic. So the models these theories provide, and derivately any situations that can be appropriately represented by them, should have causal structures that satisfy screening off. The same can not be expected of quantum models and the situations they represent.

But what does this tell us about the macroscopic world in general, and in particular about the kinds of medical, social, political or economic features of it that DAG techniques are used to study? The macroscopic world is to all appearances neither deterministic nor probabilistic. It is for the most part an unruly mess. Our job is to find methods of modelling the phenomena of immediate concern that will provide us with as much predictive power as possible. Just as with social statistics, where we realise that if we want to model the associations of quantities of interest we will have more success with probabilistic models than with deterministic ones, so too with causality: models of probabilistic causality have a far greater representational range than those restricted to deterministic causality.

This is not an unrecognised lesson. Consider very simple linear models with error terms that, following the early work of the Cowles Commission, may be used in econometrics to represent causal relations:

$$x_1 = u_1$$

$$x_2 = a_{21}x_1 + u_2$$

$$x_3 = a_{31}x_1 + a_{32}x_2 + u_3$$

The convention for these representations is that the variables written on the right-hand side in each equation are supposed to pick out a set of causal parents for the feature designated by the variable on the left-hand side. The u_is represent, in one heap, measurement error, omitted factors, and whatever truly probabilistic element there may be in the determination of an effect by its causes. This last is my concern here. Simultaneous linear equation models look deterministic (fixing the values of the variables on the right-hand side completely fixes the values of the variables on the left), but they can neverthe-

[5] *Ibid.*, p. 64.

less be used to represent genuine probabilistic causality via the 'error terms' that typically occur in them. For other matters, life is made relatively simple when we assume that the error terms are independent of each other[6] (u_i is independent of u_j, for $i \neq j$): the system is identifiable, simple estimators work well, *etc*. But when the concern is how causes operate, these independence conditions on the error terms are more or less identical to the independence conditions laid down by the Markov condition for the special case of linear models.

One major trend in econometrics has moved away from these kinds of simple across-the-board assumptions. We are told instead that we have to model the error terms in order to arrive at a representation of the processes at work in generating the probabilistic relations observed.[7] So these models need not be committed to the screening-off constraint on the operation of probabilistic causes. Violations of this constraint will show up as correlations between error terms in different equations.

The use of error terms in simultaneous equations models is not a very sensitive instrument for the representation of probabilistic causality, however. Typically there is one error term for each dependent variable – *i.e.*, for each effect – whereas what ideally is needed is one term per causal factor per effect, to represent the probabilities with which that cause 'fires' at each level to contribute to the effect. As it is (even subtracting out the alternative roles for the error terms), they still can represent only the overall net probabilistic effect of all the causes at once on a given dependent variable. A finer-grained representation of exactly how a given cause contributes to its different effects is not provided. Thus the simultaneous equation models can admit violations of the screening-off condition, but they are not a very sensitive device for modelling them.

The DAG approach also has a method that allows for the representation of purely probabilistic causality in a scheme that looks formally deterministic. To construct this kind of representation, we pretend that Nature's graph is indeed deterministic, but that some of the determining factors are not known to us nor ever would be. The set of vertices is then expanded and divided into two kinds: those which stand for known quantities, including ones that it will never be possible to know about (represented by, say, empty circles as in figure 5.1) and those that stand for quantities that we do know about (represented by, say, solid circles). It is very easy then to construct examples in which the independence condition is violated for the subgraph that includes

[6] This in fact is a really powerful motive for aiming for models with independent error terms. For attempts to find methods of inference that are robust even in the face of correlation, see for instance Toda and Phillips 1993.

[7] For an example of a case that has proved incredibly hard to model with independent errors, see Ding *et al*. 1993.

only the vertices standing for known quantities, but not for the full graph which we have introduced as a representational fiction.

The trouble with this strategy is that it tends to be methodologically costly. There would in principle be no harm to the strategy if it were kept clearly in mind that it, like the use of error terms in simultaneous equations models, is a purely representational device. But that may be far removed from what happens in practice. In a number of social science applications I have observed, quite the opposite is the case. Rather than first taking an informed case-by-case decision about whether, for the case at hand, we are likely to be encountering causes that act purely probabilistically, we instead set out on a costly and often futile hunt to fill in the empty circles. The motive may be virtuous: the DAG theorems justify a number of our standard statistical methods for drawing causal inferences *if* we can get our graphs complete enough. Attempting to fill in the graphs is a clear improvement over much social science practice that uses the familiar inference techniques without regard for whether they are justified. But the far better strategy is to fit our methods to our best assessment of the individual situations we confront. This means that we may generally have to settle for methods – like the old-fashioned hypothetico-deductive method – that provide less certainty in their conclusions than those offered by DAG proponents, but are more certainly applicable.

We began this section with the observation that the Markov condition can be assumed wherever the causal structures in our graphs are not genuinely indeterministic but only *pseudo-indeterministic*. Before closing, we should look at other reasons Spirtes, Glymour and Scheines themselves give in support of the Markov condition. These appear in the concluding two paragraphs of their section 3.5:

> The Causal Markov Condition is used all the time in laboratory, medical and engineering settings, where an unwanted or unexpected statistical dependency is *prima facie* something to be accounted for. If we give up the Condition everywhere, then a statistical dependency between treatment assignment and the value of an outcome variable will never require a causal explanation and the central idea of experimental design will vanish. No weaker principle seems generally plausible; if, for example, we were to say only that the causal parents of Y make Y independent of more remote *causes*, then we would introduce a very odd discontinuity: So long as X has the least influence on Y, X and Y are independent conditional on the parents of X. But as soon as X has no influence on Y whatsoever, X and Y may be statistically dependent conditional on the parents of Y.
>
> The basis for the Causal Markov Condition is, first, that it is necessarily true of populations of structurally alike pseudo-indeterministic systems whose exogenous variables are distributed independently, and second, it is supported by almost all of our experience with systems that can be put through repetitive processes and whose fundamental propensities can be tested. Any persuasive case against the Condition

would have to exhibit macroscopic systems for which it fails and give some powerful reason why we should think the macroscopic natural and social systems for which we wish causal explanations also fail to satisfy the condition. It seems to us that no such case has been made.[8]

I shall look at each point in turn.

How do we eliminate factors as causes when they have 'an unwanted or unexpected' correlation with the target effects of the kind I have described? In particular what kind (if any) of probabilistic test will help us tell a by-product from a cause? The kind of worry I raise is typical in medical cases, where we are very wary of the danger of confusing a co-symptom with a cause. For just one well-known example to fix the structure, recall R. A. Fisher's hypothesis that, despite the correlation between the two, smoking does not cause lung cancer. Instead, both are symptoms of a special gene.

One good strategy for investigating hypotheses like this where we are unwilling to assume that the putative joint cause must satisfy the Markov condition[9] is to conduct a randomised treatment-control experiment with the questionable factor as treatment. In an ideal experiment of this kind, we must introduce the treatment by using a cause that has no way to produce the effect other than via the treatment.[10] That means that we must not use a common cause of the effect and the treatment; so in particular we must not introduce the treatment by using a cause which has both the treatment and the target effect as by-products. If the probability of the effect in the treatment group is higher than in the control group we have good reason to think the treatment is a genuine cause and not a joint effect. This line of reasoning works whether or not the possible joint cause[11] (the gene in Fisher's example) screens off the two possible effects (smoking and lung cancer) or not. So we do not *need* to use the Markov condition in medical, engineering and laboratory settings to deal with 'unwanted' correlations. But even if we had no alternatives, it would not be a good idea to assume the principle just for the sake of having something to do. If we do use it in a given case, the confidence of our results must be clearly constrained by the strength of the evidence we have that the causes in this case do satisfy the principle.

Next, Spirtes, Glymour and Scheines claim that if we give up the Markov principle everywhere, probabilistic dependencies will not require explanation.

[8] Spirtes et al. 1993.

[9] We may be unwilling either because we know enough about genes to think the gene would produce the two effects in tandem and not independently; or we know very little, and so should not assume either dependence or independence in the production process.

[10] Thus our confidence in our conclusions can be no higher than our confidence that we have succeeded in doing this.

[11] Or, in accord with the Spirtes, Glymour and Scheines Markov condition, which holds fixed *all* the causes of the targeted effect, 'whether the possible joint cause plus the method by which the treatment is introduced screen off the two possible effects or not'.

I do not see why this claim follows, though it clearly is true that without the Markov condition, we would sometimes have different kinds of explanations. But at any rate, I do not think anyone claims that it fails everywhere.

As to the weaker condition they discuss, this is the one that I said in section 2.1 would be more natural to call a 'Markov' condition. I will argue in the next section that the ideas motivating this kind of condition may not always be appropriate when we are discussing laws, as we are here, rather than singular causal claims. But that is beside the point right now. The point here is that this weaker principle excludes causal action across temporal gaps. What Spirtes, Glymour and Scheines complain about is that it does not talk about anything else. But it does not seem odd to me that a principle should confine itself to a single topic. I admit that if all we know about how a given set of causes acts is that it satisfies this weaker principle, we will not be in a very strong position to make causal inferences. But again, that is not a reason to supplement what we may have good reason to assume about the set, by further principles, like screening off, that we have no evidence for one way or another.

Next, we see that Spirtes, Glymour and Scheines claim as their second major defence of screening off that it is supported by 'almost all of our experience with systems that can be put through repetitive processes and whose fundamental propensities can be tested'.[12] Here, it seems to me, they must be failing to take account of all those areas of knowledge where probabilistic causality is the norm, and where products and by-products, sets of multiple symptoms, and effects and side-effects are central topics of concern. Drug studies are one very obvious – and mammoth – example. We have vast numbers of clinical trials establishing with a high degree of certainty stable probabilistic relations between a drug and a variety of effects, some beneficial and some harmful; and independence among the effects conditional on the drug being used is by no means the norm.

Last, do we have 'powerful reason' for thinking the macroscopic systems we want to study fail the screening-off condition? I have given three kinds of reasons:

(1) In a great variety of areas we have very well tested claims about causal and probabilistic relations and those relations do not satisfy the screening-off condition.[13]

(2) Determinism is sufficient for screening off. But our evidence is not sufficient for universal determinism. To the contrary, for most cases of causality we know about, we do not know how to fit even a probabilistic model,

[12] *Ibid.*, p. 64.

[13] Though of course they are *consistent* with the claim that there are in fact missing (unknown) causal laws that, when added, will restore screening off.

let alone a deterministic one. The assumption of determinism is generally either a piece of metaphysics that should not be allowed to affect our scientific method, or an insufficiently warranted generalisation from certain kinds of physics and engineering models.

(3) Causes make their effects happen. Adding a cause *in the right way* should thus increase the probability of an effect. This is the start of an argument that accounts for why in an ideal randomised treatment-control experiment we expect that the probability of the effect will be higher in the treatment group only if the treatment is a cause of the effect. The argument works whether we assume causes are deterministic or only probabilistic. Similarly, it is the start of an argument that, if we condition on all the other causes of a given effect occurring at the same time (if we can make sense of this), a cause will increase the probability of its effect.

Think now not about the relations between the cause and any of its effects, but instead about the relations among the effects themselves. The cause must (in the right circumstances) increase the probability of each of its effects.[14] But how do the effects relate to each other, given that the cause occurs? As we have seen, independence is only one very special case. We need an argument that this special case is the only one possible, and we do not seem to have any. This contrasts with the discussion in the last paragraph, where the relation between a cause and its effects was at stake. There we can trace out good arguments why the relations we demand should hold.

Are these reasons powerful enough? Yes. When we use a method in science, our results can be no more secure than the premises that ground the method. Clearly for a number of causes we have very good reasons from the content of the cases themselves to assume screening off. But these three kinds of arguments should make us strongly hedge our bets if we use the assumption without any reasons special to the case.

2.3 Temporal gaps and gaps in probabilities

There is a very different line of argument that supports a modified version of the screening-off condition. The argument relies on the requirement that causes must be spatially and temporally connected with their effects. The assumption that causes cannot produce effects across a gap in space or time has a very long tradition; it is in particular one of the conditions that David Hume imposed on the cause-effect relation. The argument from this requirement to a modified independence condition goes like this. Imagine that a particular cause, C, operates at a time t to produce two effects, X and Y, in correlation, effects that occur each at some later time and some distance

[14] Or, at least we have an argument to support this.

away, where we have represented this on a DAG with *C* as the causal parent of *X* and *Y*. We know there must be some continuous causal process that connects *C* with *X* and one that connects *C* with *Y*, and the state of those processes at any later time t_2 must contain all the information from *C* that is relevant at that time about *X* and *Y*. Call these states P_X and P_Y. We are then justified in drawing a more refined graph in which P_X and P_Y appear as the parents of *X* and *Y*, and on this graph the independence condition will be satisfied for *X* and for *Y* (although not for P_X and P_Y). Generalising this line of argument we conclude that any time a set of factors on an accurate graph does not satisfy the independence condition, it is possible to embed that graph into another accurate graph that does satisfy independence for that set of factors.

What is wrong with this argument is that it confuses singular causal claims about individual events that occur at specific times and places (*e.g.*, 'My taking two aspirins at 11:00 this morning produced headache relief by noon') with general claims about causal relations between *kinds* of events (*e.g.*, 'Taking aspirin relieves headaches an hour later') – claims of the kind we identify with scientific laws. Simultaneous equation models, DAGs, and path diagrams are all scientific methods designed for the representation of generic-level claims; they are supposed to represent causal laws (which are relations between kinds), not singular causal relations (which hold between individual events). Arguments that convince us that there must be a continuous process to connect every individual causal event with its distant effect event do not automatically double as arguments to establish that we can always put more vertices between causes and effects in a graph that represents causal laws.

Although it is necessary for the causal message to be transmitted somehow or other from each individual occurrence of a cause-kind to the individual occurrence of its effect-kind, the processes used to carry the message can be highly varied and may have nothing essential in common. The transmission of the causal message between the given instance of a cause-kind and the associated instance of the effect-kind can piggy-back on almost anything that follows the right spatio-temporal route and has an appropriate structure. For some causal laws the cause itself may initiate the connecting process, and in a regular law-like way. For these laws there will be intermediate vertices on a more refined graph. But this is by no means necessary, nor I think, even typical.

In principle all that is needed is to have around enough processes following the space-time routes that can carry the right sort of 'message' when the causes operate. The connection of these processes with either the general cause-kind or the general effect-kind need have none of the regularity that is necessary for a law-claim. Glenn Shafer makes a similar point in his defence of the use of probability trees over approaches that use stochastic processes. As Shafer says, 'Experience teaches us that regularity can dissolve into irregularity when we insist on making our questions too precise, and this lesson applies in particular when the desired precision concerns the timing of cause and effect.'[15]

This point is reinforced when we realise that the kind of physical and institutional structures that guarantee the capacity of a cause to bring about its effect may be totally different from those that guarantee that the causal message is transmitted. Here is an example I will discuss at greater length in chapter 7. My signing a cheque at one time and place drawn on the Royal Bank of Scotland in your favour can cause you to be given cash by your bank at some different place and time, and events like the first do regularly produce events like the second. There is a law-like regular association and that association is a consequence of a causal capacity generated by an elaborate banking and legal code. Of course the association could not obtain if it were not regularly possible to get the cheque from one place to the other. But it is; and the ways to do so are indefinitely varied and the existence of each of them depends on quite different and possibly unconnected institutional and physical systems: post offices, bus companies, trains, public streets to walk along, legal systems that guarantee the right to enter the neighbourhood where the bank is situated, and so on and so on.

The causal laws arising from the banking system piggy-back on the vast number of totally different causal processes enabled by other institutions, and there is nothing about those other processes that could appropriately appear as the effect of cheque-signing or the parent of drawing-cash-on-the-cheque. We might of course put in a vertex labelled 'something or other is happening that will ensure that the signed cheque will arrive at the bank', but to do that is to abandon one of the chief aims we started out with. What we wanted was to establish causal relations by using information that is far more accessible, information about readily identifiable event-types or easily measured quantities and their statistical associations. Here we have fallen back on a variable that has no real empirical content and no measurement procedures associated with it.

2.4 *Aside: the Faithfulness condition*

The arguments here clearly do not oppose the Markov condition *tout court*. They show rather that it is not a universal condition that can be imposed willy-nilly on all causal structures. The same is true for a second important condition that DAG-techniques employ: the Faithfulness condition. This condition is roughly the converse of screening off. Screening off says that, once the parents of a factor have been conditioned on, that factor will be independ-

[15] Shafer 1997, p. 6.

ent of everything except its effects; *i.e.*, probabilistic dependence implies causal dependence. The Faithfulness condition supposes that probabilistic dependencies will faithfully reveal causal connections; *i.e.*, causal dependence implies probabilistic dependence.

The standard argument against Faithfulness points to causes that act in contrary ways. Via one route a cause prevents the effect; by another, it causes it. Faithfulness will be violated if the two processes are equally effective and cancel each other out. It is not uncommon for advocates of DAG-techniques to argue that cases of cancellation will be extremely rare, rare enough to count as non-existent. That seems to me unlikely, both in the engineered devices that are sometimes used to illustrate the techniques and in the socio-economic and medical cases to which we hope to apply the techniques. For these are cases where means are adjusted to ends and where unwanted side effects tend to be eliminated wherever possible, either by following an explicit plan or by less systematic fiddling. The case of solitons, discussed in chapter 1, section 4, provides a good illustration. There the natural Doppler broadening is just offset by the non-linear chirp effects in order to produce a stable pulse. The bad effects of a feature we want – or are stuck with – are offset by enhancing and encouraging its good effects. Whether we do it consciously or unconsciously, violating the Faithfulness condition is one of the ways we minimise damage in our social systems and in our medical regimens.

As with the Markov condition, I would not wish to argue that the Faithfulness condition never holds. That would be as much a mistake as to argue that it always holds. To decide whether it does or not requires having some understanding of how the causal structure under study works, and any conclusions we draw about causal relations using methods that presuppose Faithfulness can only be as secure as our models of that structure and its operation. That is why I urge 'No models in, no causes out.'

2.5 *The many kinds of causation*

I have argued that neither the Markov nor the Faithfulness condition can serve as weapons in an arsenal of universal procedures for determining whether one factor causes another. I should like to close with some brief consideration of why this is so. The mistake, it seems to me, is to think that there is any such thing as the causal relationship for which we could provide a set of search procedures. Put that way, this may seem like a Humean point. But I mean the emphasis to be on the *the* in the expression 'no such thing as the causal relationship'. There are such things as causal relations, hundreds of thousands of them going on around us all the time. We can, if we are sufficiently lucky and sufficiently clever, find out about them, and statistics can play a crucial role in our methods for doing so. But each causal relation may be different from each other, and each test must be made to order.

Consider two causal hypotheses I have worked with. The first is the principle of the laser: sparking an inverted population stimulates emission of highly coherent radiation. The second is an important principle we have looked at in chapter 1, the principle that makes fibre bundles useful for communication: correct matching of frequency to materials in a fibre bundle narrows the Doppler broadening in packets travelling down the fibre. In both cases we may put the claims more abstractly, with the causal commitment made explicit: 'Sparking an inverted population causes coherent radiation' and 'Correct frequency matching causes wave-packets to retain their shapes.' Now we have two claims of the familiar form, 'X causes Y.' But no one would take this to mean that the same relation holds in both cases or that the same tests can be applied in both cases. Yet that is just the assumption that most of our statistical methodology rests on, both methodologies that use DAG-techniques and methodologies that do not.

I have argued against the assumption that we are likely to find universal statistical methods that can decide for us whether one factor causes another. But we must not follow on from that to the stronger view that opponents of causal interference often take, that we cannot use statistics at all to test specific causal hypotheses. We all admit the truism that real science is difficult; and we are not likely to find any universal template to carry around from one specific hypothesis to another to make the job easy. In so far as DAG methods purport to do that, it is no surprise that they might fail. Yet with care and caution, specialists with detailed subject-specific knowledge can and do devise reliable tests for specific causal hypotheses using not a universally applicable statistical method but rather a variety of statistical methods – DAG techniques prominent among them – different ones, to be used in different ways in different circumstances, entirely variable from specific case to specific case. Science is difficult; but it has not so far proved to be impossible.

If, as I claim, there is no such thing as *the* causal relation, what are we to make of claims of the form '*X* causes *Y*' – *e.g.*, 'Correct frequency matching causes wave-packets to retain their shape'? Consider first, what do we mean by 'cause' here? The *Concise Oxford English Dictionary* tell us that 'to cause' is 'to effect', 'to bring about' or 'to produce'. Causes make their effects happen. That is more than, and different from, mere association. But it need not be one single different thing. One factor can contribute to the production or prevention of another in a great variety of ways. There are standing conditions, auxiliary conditions, precipitating conditions, agents, interventions, contraventions, modifications, contributory factors, enhancements, inhibitions, factors that raise the number of effects, factors that only raise the level, *etc*.

But it is not just this baffling array of causal roles that is responsible for the difficulties with testing – it is even more importantly the fact that the term 'cause' is abstract. It is abstract relative to our more humdrum action verbs in just the sense introduced in chapter 2: whenever it is true that '*X* causes *Y*', there will always be some further more concrete description that the causing consists in. This makes claims with the term 'cause' in them *unspecific*: being abstract, the term does not specify the form that the action in question takes. The cat causes the milk to disappear; it *laps it up*. Bombarding the population of atoms of a ruby-rod with light from intense flash lamps causes an inversion of the population; it *pumps* the population to an inverted state. Competition causes innovation. It *raises* the number of patents; it *lowers* production costs.

This matters for two reasons. The first is testing. Reliable tests for whether one factor causes another must be finely tuned to how the cause is supposed to function. With fibre bundles, for example, we proceed dramatically differently for one specification, say, 'Correct matching gates the packets', from the different causal hypothesis, 'Correct matching narrows the packets.' The two different hypotheses require us to look for stability of correlation across different ranges of frequencies and materials, so the kinds of statistical tests appropriate for the first are very different from the kinds appropriate for the second. The second reason is policy. We may be told that *X* causes *Y*. But what is to be done with that information is quite different if, for example, *X* is a precipitating cause from what we do if it is a standing cause, or an active cause versus the absence of an impediment.

Here is a homely analogue to illustrate the problems we have both in testing claims that are unspecific and in using them to set policy. I overheard my younger daughter Sophie urging her older sister Emily to turn off the television: 'Term has just begun and you know mama is always very irritable when she has to start back to teaching.' Sophie's claim, 'mama is irritable', was true but unspecific. It provided both a general explanation for a lot of my peculiar behaviour of late and also general advice: 'Watch your step.' But Emily wanted to know more. 'Don't you think I can get away with just another half hour on the telly?' Sophie's hypothesis could not answer. Nor, if called upon, could she test it. Behaviourist attempts to operationalise irritability failed not just because mental states are not reducible to patterns of behaviour but equally because descriptions like *irritable* are unspecific. Claims of the form '*X* causes *Y*' are difficult to test for the same reason. Sophie's claim is also, like the general claim '*X* causes *Y*', weak in the advice-giving department. It gives us a general description of what to do,[16]

[16] The general advice that follows from '*X* causes *Y*' seems to be of this form: if you want *Y*, increase the frequency (or level) of *X* without in any way thereby increasing the frequency

but it does not tell us if we can risk another half hour on the telly. Usually in looking for causal information we are in Emily's position – we are concerned with a very concrete plan of action. So then we had better not be looking to claims of the form '*X* causes *Y*' but instead for something far more specific.

3 Causal structures and the nomological machines that give rise to them

3.1 Where do causal structures come from?

Following the terminology that is now standard, I spoke in the last section about causal structures. What is a causal structure; and where does it come from? According to Spirtes, Glymour and Scheines, a causal structure is

> an ordered pair <*V,E*> where *V* is a set of variables, and *E* is a set of ordered pairs of *V*, where <*X,Y*> is in *E* if and only if *X* is a direct cause of *Y* relative to *V*.[17]

Alternatively, *V* can be a set of events. But we should not be misled into thinking we are talking about specific events occurring at particular times and places. The causal relations are supposed to be nomic relations between event-types, not singular relations between tokens.

Where does the complex of causal laws represented in a causal structure come from and what assures its survival? These laws, I maintain, like all laws, whether causal or associational, probabilistic or deterministic, are transitory and epiphenomenal. They arise from – and exist only relative to – a nomological machine. In figure 5.2 we have a typical graph of the kind Spirtes, Glymour and Scheines employ to represent sets of causal laws. What we see represented in the graph are fixed connections between event-types. One event-type either causes another or it does not, and it does so with a definite and fixed strength of influence if Spirtes, Glymour and Scheines' best theorems are to be applicable. We typically apply these methods to event-types like unemployment and inflation, income and parents' education level, divorce rates and mobility, or a reduction in tax on 'green' petrol and the incidence of lung cancer. What could possibly guarantee the required kind of determinate relations between event-types of these kinds? You could think of reducing everything ultimately to physics in the hope that in fundamental physics unconditional laws of the usual kind expressed in generalisations will

(or level) of features that prevent *Y*. Even that advice does not seem correct, however, if *X* is an enhancer and the basic process to be enhanced does not occur, or if *X* needs some imperfection to precipitate it and we have bought ourselves pure samples of *X* in the expectation that we will thus maximise its efficacy.

[17] Spirtes et al. 1993, p. 45.

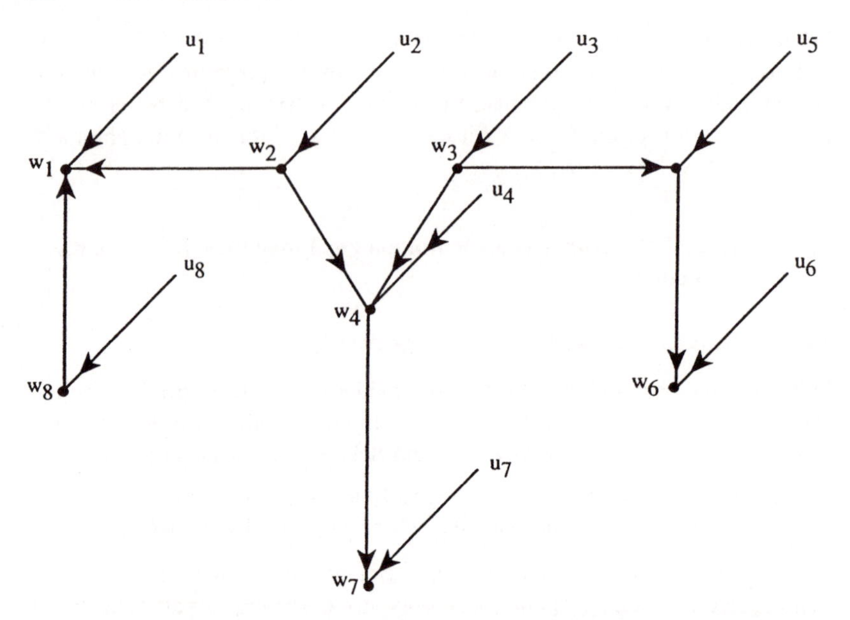

Figure 5.2 A causal structure. Source: designed by Towfic Shomar.

at last be found: proper causal laws and laws of association with modal operators and universal (or probabilistic) quantifiers in front. Apart from the implausibility of this downward reduction, this hope, I have been arguing in previous chapters, is bound to be dashed. All the way down, laws are transitory and epiphenomenal, not eternal.

I ask 'Where do causal structures come from?' I will tell you exactly how the particular structure pictured in figure 5.2 was generated. It comes from the machine in figure 5.3. This machine is a nomological machine: it has fixed components with known capacities arranged in a stable configuration. It is the fixed arrangement among the parts that guarantees the causal relations depicted in the graph. (All the machines figured in this chapter plus the associated causal graphs have been designed by Towfic Shomar.) The claim I have been arguing is that we always need a machine like this to get laws – any laws, causal or otherwise. Sometimes God supplies the arrangements – as in the planetary systems – but very often we must supply them ourselves, in courtrooms and churches, institutions and factories. Where they do not exist, there is no sense in trying to pick out event-types and asking about their nomological relations. That would be like asking: 'Given we drop six balls in a bucket, what is the probability that a second bucket ten feet away will rise by six inches?' The question makes no sense unless we have in mind some background machine to which the buckets are bolted.

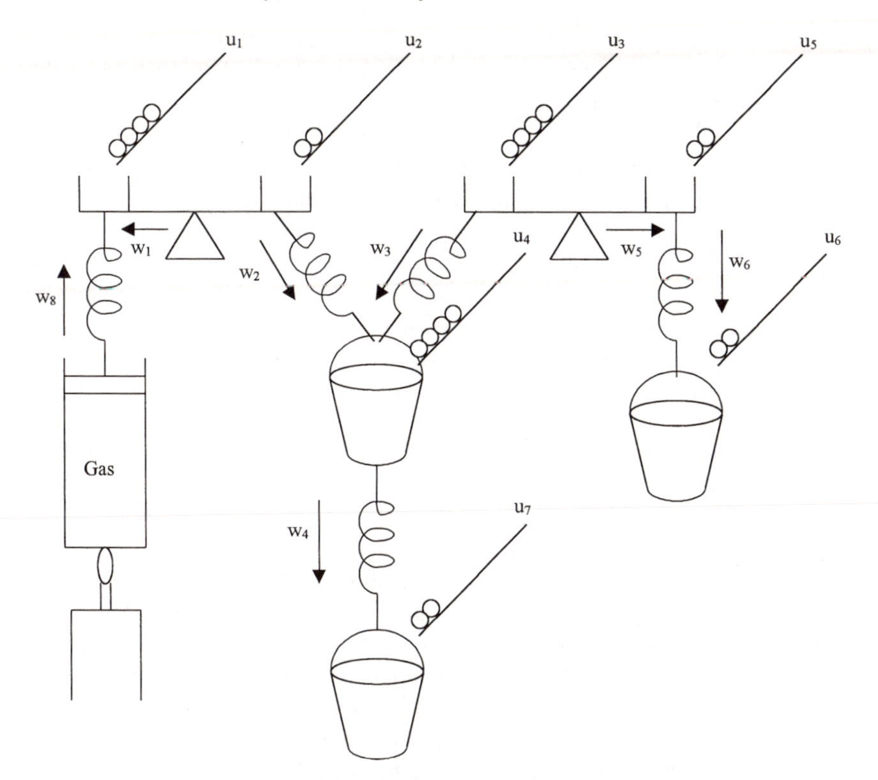

Figure 5.3 The nomological machine giving rise to the structure of Figure 5.2. Source: designed by Towfic Shomar; recreation, George Zouros.

It is interesting to note that this way of thinking about the two levels is similar to the way matters were originally viewed when the kinds of probability methods we are now studying were originally introduced into social enquiry. The founders of econometrics, Trygve Haavelmo and Ragnar Frisch, are good examples. Both explicitly believed in the socio-economic machine. Frisch for instance proposed a massive reorganisation of the economy relying on the economic knowledge that he was acquiring with his new statistical techniques.[18] Or consider Haavelmo's remarks about the relation between pressure on the throttle and the acceleration of the car.[19] This is a perfectly useful piece of information if you want to drive the car you have, but it is not what you need to know if you are expecting change. For that you need to understand how the fundamental mechanisms operate.

[18] *Cf.* Andvig 1988.
[19] Haavelmo 1944.

My own hero Otto Neurath expressed the view clearly before the First World War in criticising conventional economics. Standard economics, he insisted, makes too much of the inductive method, taking the generalisations that hold in a free market economy to be generalisations that hold *simpliciter*: 'Those who stay exclusively with the present will very soon only be able to understand the past.'[20] Just as the science of mechanics provides the builder of machines with information about machines that have never been constructed, so too the social sciences can supply the social engineer with information about economic orders that have never yet been realised. The idea is that we must learn about the basic capacities of the components; then we can arrange them to elicit the regularities we want to see. The causal laws we live under are a consequence – conscious or not – of the socio-economic machine that we have constructed.

In the remainder of this chapter I want to make a small advance in favour of the two-tiered ontology I propose. I will do this by defending a distinction between questions about what reliably causes what in a given context and questions about the persistence of this reliability. The first set of questions concerns a causal structure; the second is about the stability of the structure. Recall the machine in figure 5.3 that supports the graph in figure 5.2. There is a sense in which this is a 'badly shielded' machine. Normally we want to construct our machine so that it takes inputs at very few points. But for this machine every single component is driven not only by the other components of the machine but also by influences from outside the system. We had to build it like that to replicate the graphs used by Spirtes, Glymour and Scheines. In another sense, though, the machine is very well shielded, for nothing disturbs the arrangement of its parts.

My defence here of the distinction between the causal structure and the nomological machine that gives rise to it will consist of two separate parts: the first is a summary of a talk I gave with Kevin Hoover[21] at the British Society for the Philosophy of Science (BSPS) in September 1991, and the second is a local exercise to show the many ways in which this distinction is maintained in the Spirtes, Glymour and Scheines work. In particular I try to show why it would be a mistake to let this distinction collapse by trying to turn the nomological machine into just another kind of cause.

3.2 *Stability*[22]

My discussion with Kevin Hoover focused on Mary Morgan's remark that the early econometricians believed that their statistical techniques provided *a*

[20] Quoted from Nemeth 1981, p. 51, trans. by Nancy Cartwright.
[21] For Kevin Hoover's views on causality, see Hoover forthcoming.
[22] For a further discussion of experiments and stability see Cartwright 1995d and 1995e.

substitute for the experimental method. I defended their belief. As I see it, the early econometrics work provided a set of rules for causal inference (RCI):

RCI: Probabilities and old causal laws → New causal laws

Why should we think that these rules are *good* rules? The econometricians themselves – especially Herbert Simon – give a variety of reasons; I gave a somewhat untidy account in *Nature's Capacities and their Measurement.* At the BSPS I reported on a much simpler and neater argument that is a development of Hoover's work (following Simon) and my earlier proof woven together. This was the thesis:

Claim: RCI will lead us to infer a causal law from a population probability only if that law would be established in an ideal controlled experiment in that population.

That is, when the conditions for inferring a causal law using RCI are satisfied, we can read what results would obtain in an experiment directly from the population probabilities. We do not need to perform the experiment – Nature is running it for us. That is stage 1. The next stage is an objection that Hoover and I both raise to our arguments. The objection can be very simply explained without going into what the RCI really say. It is strictly analogous to a transparent objection to a very stripped-down version of the argument. Consider a model with only two possible event types, C_t and $E_{t+\Delta t}$. Does C at t cause E at $t + \Delta t$? The conventional answer is that it does so if and only if C_t raises the probability of $E_{t+\Delta t}$, *i.e.* (suppressing the time indices),

$$RCI_{2\text{-event model}}: C \text{ causes } E \text{ iff } Prob(E/C) > Prob(E/-C)$$

The argument in defence of this begins with the trivial expansion:

$$Prob(E) = Prob(E/C) \cdot Prob(C) + Prob(E/-C) \cdot Prob(-C)$$

Let us now adopt, as a reasonable test, the assumption that if C causes E then (*ceteris paribus*) raising the number of Cs should raise the number of Es. In the two-variable model there is nothing else to go wrong, so we can drop the *ceteris paribus* clause. Looking at the expansion, we see that Prob(E) will go up as Prob(C) does if and only if Prob(E/C) > Prob($E/-C$). So our mini-rule of inference seems to be vindicated.

But is it really? That depends on how we envisage proceeding. We could go to a population and collect instances of Cs. If Prob(E/C) > Prob($E/-C$), this will be more effective in getting Es than picking individuals at random. But this is not always what we have in mind. Perhaps we want to know whether a *change* in the level of C will be reflected in a change in the level of E. What happens then? Suppose we propose to effect a change in the level of C by manipulating a 'control' variable W (which for simplicity's sake is a 'switch' with values $+W$ or $-W$). Now we have three variables and it seems we must consider a probability over an enlarged event space:

Prob($\pm E \pm C \pm W$). Once W is added in this way, the joint probability of C and E becomes a conditional probability, conditional on the value of W. It look as if the Prob(E) that we have been looking at then is really Prob(E/$-W$). But what we want to learn about is Prob(E/W), and the expansion

$$\begin{aligned}\text{Prob}(E/-W) = {} & \text{Prob}(E/C-W)\cdot\text{Prob}(C/-W) \\ & + \text{Prob}(E/-C-W)\cdot\text{Prob}(-C/-W)\end{aligned}$$

will not help, since from the probabilistic point of view, there is no reason to assume that

$$\text{Prob}(E/C+W) = \text{Prob}(E/C-W)$$

But probabilities are not all there is. In my view probabilities are a shadow of the underlying machine. The question of which are fundamental, nomological machines or probabilities, matters because reasoning about nomological machines can give us a purchase on invariance that probabilities cannot provide. To see this let us look once again at the Markov condition.

Think again of our model (C,E,W) in terms of just its causal structure. The hypothesis we are concerned with is whether C at t causes E at $t + \Delta t$ in a toy model with very limited possibilities: (i) W at some earlier time t_0 is the complete and sole cause of C at t; and (ii) C is the only candidate cause for E at t. In this case the invariance we need is a simple consequence of the Markov assumption: a full set of intermediate causes screens off earlier causes from later effects. (This is the part of the condition that seems to me appropriately called 'Markov'.) In our case this amounts to

$$\text{MC: Prob}(E/\pm C \pm W) = \text{Prob}(E/\pm C)$$

Spirtes, Glymour and Scheines have a theorem to the same effect, except that their theorem is geared to their graph representation of complex sets of causal laws. Kevin Hoover and I use the less general but more finely descriptive linear-modelling representations. Spirtes, Glymour and Scheines call theirs the 'manipulation theorem'. They say of it,

> The importance of this theorem is that if the causal structure and the direct effects of the manipulation . . . are known, the joint distribution [in the manipulated population] can be estimated from the unmanipulated population.[23]

Let us look just at the opening phrases of the Spirtes, Glymour and Scheines theorem:

[23] Spirtes et al. 1993, p. 79.

Given directed acyclic graph G_{Comb} over vertex set $\mathbf{V} \cup \mathbf{W}$ and distribution $P(\mathbf{V} \cup \mathbf{W})$ *that satisfies the Markov condition* for G_{Comb}, . . .[24]

So the Markov condition plays a central role in the Spirtes, Glymour and Scheines argument as well.

What exactly are we presupposing when we adopt condition MC? Two different things I think. The first is a metaphysical assumption that in the kinds of cases under consideration causes do not operate across time gaps; there is always something in between that 'carries' the influence. Conditioning on these intermediate events in the causal process will render information about the earlier events irrelevant, so the probability of E given C should be the same regardless of the earlier state of W.

I objected to this assumption about conditioning in section 2.3 on the grounds that the intermediate factors that ensure temporal continuity for the individual causes may not fall under any general kinds of the right type and hence may not appear in the causal structure, since it represents causal laws not singular causal facts. But more is involved. Clearly the assumption that $\text{Prob}(E/\pm C)$ is the same given either $+W$ or $-W$ assumes that changes in W are not associated with changes in the underlying nomological machine. This is required not just with respect to the qualitative causal relation – whether or not C causes E – but also with respect to the exact quantitative strengths of the influence, which are measured by the conditional probabilities (or functions of them). This assumption is built into the Spirtes, Glymour and Scheines theorem in two ways: (1) The antecedent of the theorem restricts consideration to distributions $\text{Prob}(\mathbf{V} \cup \mathbf{W})$ that satisfy the Markov condition, a restriction which, as we will see in the next section, has no general justification in cases where causal structures vary; (2) they assume in the proof that $G_{Unman} = G_{Man}$ are subgraphs of G_{Comb}. That is, they assume that changes in the distribution of the externally manipulated 'switch variable' W are not associated with any changes in the underlying causal structure. That is the same assumption that Hoover and I pointed to as well.

Where do we stand if we build this assumption in, as Spirtes, Glymour and Scheines do? It looks as if the original thesis is correct: *if* changes in the variables that serve as control variables in Nature's experiment are not associated with changes in the remainder of the causal structure *then* RCI (or the Spirtes, Glymour and Scheines programme) will allow us to infer causal laws from laws of probabilistic association. But it is the nomological machine that makes the structure what it is. The way we switch W on and off must not interfere with that if we are to expect the laws it gives rise to be the same once we have turned W on. So the RCI of our miniature 2-variable model

[24] *Ibid.*, italics added.

are good rules of inference – but only if the underlying machine remains intact as W and C change. That seems to be the situation with respect to the back and forth about causal inference between Hoover and me at the BSPS meeting. The reason for covering this again is to point out that central to the argument is the distinction between (1) the nomological machine that gives rise to the laws and (2) the laws themselves – both the causal laws and the associated laws of probabilistic association.

3.3 Causal auxiliaries versus causal structures

There is an obvious strategy for trying to undo the distinction between (1) and (2). That is to try to take the nomological machine 'up' from its foundational location and to put it into each of the emerging causal laws themselves, on the model of an auxiliary. Following John Stuart Mill[25] or John L. Mackie[26] (as Spirtes, Glymour and Scheines do too) suppose we take the law connecting causal parents with their effects to have the form:

$$C_1A_1 \vee C_2A_2 \vee \ldots \vee C_nA_n \text{ causes } E,$$

where the Cs are the 'salient' factors and the As the necessary helping conditions that make up a complete cause. The idea is to include a description that picks out the nomological machine that gives rise to the causal relation between the C_i and E in each of the disjuncts. That is, to assume for each A_i in the above formula that

$$A_i \supset NM = x$$

where NM is a random variable whose values represent the range of possible structures and x is the particular value that NM is supposed to take when just the machine in question obtains. In the case of the causal structure in figure 5.2, for example, NM must take a value which singles out just the machine in figure 5.3. There are a number of objections to this strategy.

First, the proposal is obviously impracticable. Just think of it narrowly from the point of view of the Spirtes, Glymour and Scheines framework. Suppose we are studying a set of causally sufficient variables (so that every common cause of every member of the vertex set is in the vertex set). This is the case for which they get the most powerful theorems. If the description of the causal structure must be included at this level, the set would be radically insufficient and we could, artificially, no longer be entitled to the results of those theorems.

Second is the very idea of this new random variable. Its values must repre-

[25] Mill 1843.
[26] Mackie 1965.

sent all possible nomological machines that could give rise to some causal structure or other over the designated set of vertices. Does this even make sense? And what kinds of values should these be (integers, real numbers, complex numbers, vectors . . .?) and what will make for an ordering on them? For an ordinary random variable of the kind that usually appears in the vertex set of a causal structure, we expect that there should be a fairly uniform set of fairly specific procedures for measuring the variable, regardless of its value. That does not seem possible here. We also expect that the order relation on the values reflects some ordering of magnitudes represented by those values. Distances too are generally significant: if the distance between one pair of values is greater than between a second pair, we expect two items represented by the first pair to differ from each other in some respect in question more than two items represented by the second pair.

These are just the simplest requirements. In general when we construct a new random variable, we first have in mind some features that the qualitative relations that items to be assigned values have. For temperature, for example, the items can be completely ordered according to '. . . is hotter than . . .'. Then we construct the right kind of mathematical object to represent those features. And when we do it formally, we finally try to prove a representation theorem, to show that the two match in the intended way. None of this seems to make sense for the random variable labelled '*NM*'. It seems we are trying to press nomological machines into the same mould as the measurable quantities that appear as causes and effects, but they do not fit.

This leads immediately to a third objection: nomological machines are not easy to represent in the same way that we represent causes because they aren't causes. Look again at our paradigm, Mackie's and Mill's account. The relationship between the *C*s and *A*s on the one hand and the *E*s on the other is a causal relation. That is not true of the relation between the nomological machine and the set of laws (causal and probabilistic) that it gives rise to. Consider the machine of figure 5.3 and its concomitant causal graph. The machine causes neither the causal laws in the graph nor the effects singled out in the causal laws pictured there. The importance of this is not just metaphysical; there are real methodological implications. The two types of relation are very different. So too will be the methods for investigating them. In particular we must not think that questions about the stability of causal laws can be established in the same way that the laws themselves are established, for instance by including a variable that describes the causal machine in the vertex set and looking at the probabilistic relations over the bigger vertex set.

One last objection concerns how the nomological machine variable would actually have to be deployed. If it is to be guaranteed to do its job of ensuring that the Markov condition holds for a causal structure, it will have to appear as a cause of every effect in the structure. Besides being cumbersome, this can not be right. For the machine that is responsible for the causal laws is not part of the cause of the laws' effects: it is the causes cited in those laws that (quoting the *Concise Oxford Dictionary* again) 'bring about' or 'produce' the effects. Again, the difference will come out methodologically. For many causal laws in nice situations we will be able to devise statistical tests. But the same kind of tests do not make sense when the nomological machine is considered as a cause. Consider the most simple tests in a 2-variable model. Are increased levels of the putative cause associated with increased levels of the effect? For this we need a cause that has at least the first requirements of a quantity – it comes in more and less.

What about the test that looks for an increased probability of the effect when the putative cause is present compared with when it is absent? This test can not apply either, for the reasons I urge in chapter 7. There is no probability measure to be assigned across different nomological machines. In general there is no answer to the question, 'What is the probability that Towfic Shomar builds this machine rather than that machine or some other?' just as there need be no probability for one socio-economic set-up rather then another to obtain in a developing country.[27] We see thus that there are a number of defects to the proposal to escalate the nomological machine into the causal structure itself. What I want to do in the next section is to look – very locally – at how these defects are bad for the Spirtes, Glymour and Scheines techniques themselves.

3.4 Mixing

Almost all of the Spirtes, Glymour and Scheines theorems presuppose that Nature's graph – that is, a graph of the true causal relations among a causally sufficient set that includes all the variables under study – satisfy their Markov condition. But that assumption will not in general be true if the vertex set includes variables, like the nomological machine variable, whose different values affect the causal relations between other variables. The problem is one that Spirtes, Glymour and Scheines themselves discuss, although their aim is to establish causal relations not to establish stability, so they do not put it in exactly this context. Here is the difficulty: consider two different values (say $NM = 1$, $NM = 2$) for the nomological machine variable. There are cases in which the true causal relations among a set of variables V (not including NM) will satisfy the Markov condition both when $NM = 1$ and when $NM = 2$; but when they are put together, the combined graph does not. This is an instance

[27] Except of course as a question about the probability of drawing one set-up rather than another from a given sample, which is not the kind of probability that we need to match a causal structure.

Figure 5.4 A nomological machine giving rise to different causal structures.

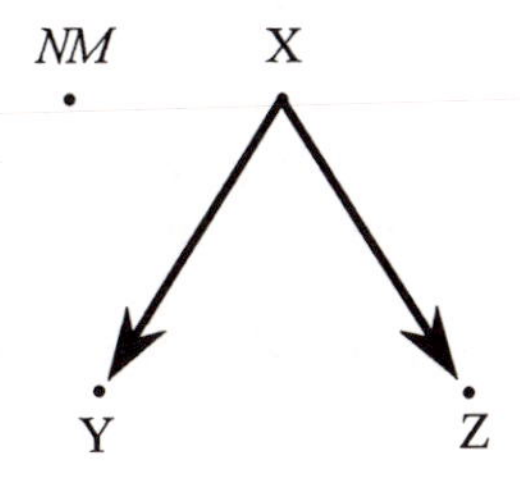

Prob(*NM*, *X*, *Y*, *Z*)

Figure 5.5 A causal structure with '*NM*' as additional variable.

of the by-now familiar Simpson's paradox. Spirtes, Glymour and Scheines call this 'mixing'.[28]

Take a simple case with two extremes, as in figure 5.4. When $NM = 1$, X causes both Y and Z deterministically. When $NM = 2$, X, Y and Z are all causally independent of each other. For Spirtes, Glymour and Scheines a causal connection appears in the graph just in case it is in the true causal relations for *any* values of any of the variables. The graph over NM, X, Y, Z is then as in figure 5.5.

The probability Prob relates to Prob_1 and Prob_2 thus:

$$\text{Prob}(\pm X \pm Y \pm Z\ /NM = 1) = \text{Prob}_1(\pm X \pm Y \pm Z)$$

$$\text{Prob}(\pm X \pm Y \pm Z\ /NM = 2) = \text{Prob}_2(\pm X \pm Y \pm Z)$$

[28] In the published version of *Causality, Prediction and Search*, they agree with me in seeing mixing as a case of Simpson's paradox though earlier drafts treated the two as distinct.

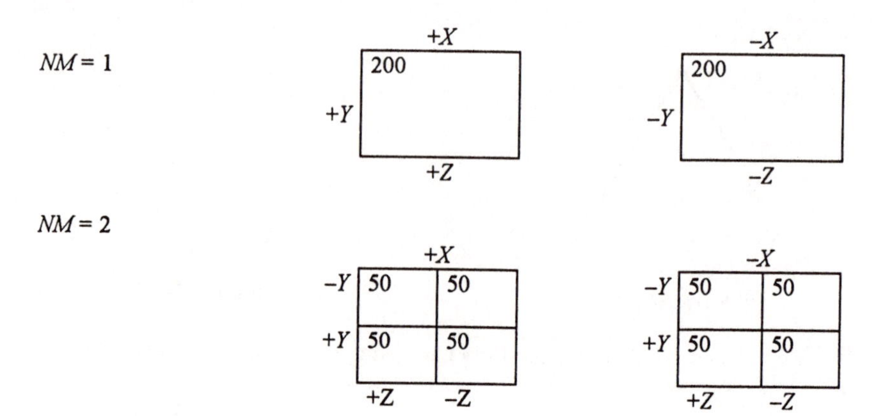

Figure 5.6 A numerical example.

The numerical example of figure 5.6 shows that both Prob_1 and Prob_2 can satisfy the Markov condition relative to their graphs, but Prob does not do so relative to its graph.

Consider Prob(+*Y*+*Z*/+*X*) from figure 5.6; the others follow the same pattern.

$$\text{Prob}_1(+Y+Z/+X) = 1 = 1 \cdot 1 = \text{Prob}_1(+Y/+X)\text{Prob}_1(+Z/+X)$$

$$\text{Prob}_2(+Y+Z/+X) = \tfrac{50}{200} = \tfrac{1}{4} = \tfrac{100}{200} \cdot \tfrac{100}{200}$$

$$= \text{Prob}_2(+Y/+X)\text{Prob}_2(+Z/+X)$$

But,

$$\text{Prob}(+Y+Z/+X) = \tfrac{250}{400} = \tfrac{5}{8} \neq \tfrac{9}{16} = \tfrac{300}{400} \cdot \tfrac{300}{400}$$

$$= \text{Prob}(+Y/+X)\text{Prob}(+Z/+X)$$

Simpson's paradox is the name given to the following fact (or its generalisation to cases with more variables and more compartments in the partition):

There are probability distributions such that a conditional dependence relation (positive, negative or zero) that holds between two variables (here *Y*, *Z*) may be changed to any other in both compartments of a partition along a third variable (here *NM*).

Defining $\text{Prob}_{\pm X}(NM \cdot Y \cdot Z) = \text{Prob}(NM \cdot Y \cdot Z/\pm X)$, we see that the example just cited exhibits Simpson's paradox for both the distribution Prob_{+X} and for Prob_{-X}. A similar construction can be made for any of the conditional independences associated with the Markov condition. A necessary condition for Simpson's paradox to obtain is that the third variable (*NM*) be conditionally

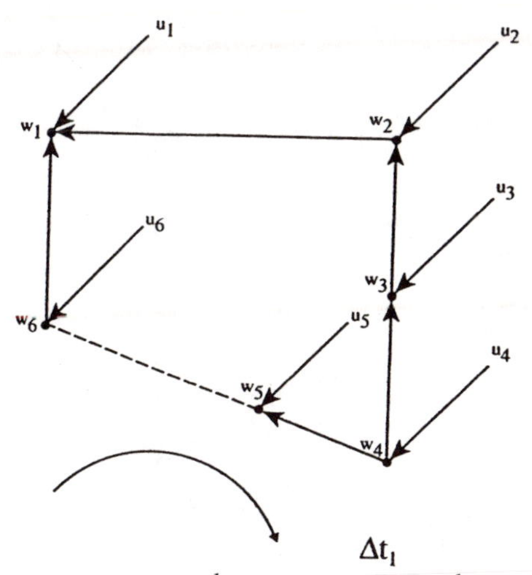

In this structure, w_2 causes w_1 and w_4 causes w_5; w_5 and w_6 are uncorrelated.

Figure 5.7a. Source: Towfic Shomar.

dependent on each of the other two (*Y*, *Z*). That is the case for each of the distributions Prob_{+X}, Prob_{-X}:

$$\text{Prob}_{+X}(Y/NM = 1) = 1 \neq 0.5 = \text{Prob}_{+X}(Y/NM = 2),$$

$$\text{Prob}_{-X}(Y/NM = 1) = 0 \neq 0.5 = \text{Prob}_{-X}(Y/NM = 2),$$

and similarly for $\text{Prob}_{\pm X}(Z/NM)$. The dependence between *Y* (or *Z*) and *NM* arises in Prob_{+X} and Prob_{-X} because, given either +*X* or –*X*, information about whether we are in *NM* = 1 or *NM* = 2 (*i.e.*, whether we are considering a population in which *X* causes *Y* (or *Z*) or not) will certainly affect the probability of finding *Y* (or *Z*).

The case I have considered here is an all-or-nothing affair: the causal law is either there and deterministic, or it is altogether missing. Figures 5.7 and 5.8 show another case where the causal graph gets entirely reversed, and not even because of changes in the structure of the machine but only because of changes in the relative values of the independent variables. But we do not need such dramatic changes. Any change in the numerical strength of the influence of *X* on *Y* and *Z* can equally generate a Simpson's paradox, and hence a failure of the Markov condition for the combined graph.[29]

[29] As an aside, let me make two remarks about the Spirtes, Glymour and Scheines scheme. First, it is important to note that mixing is a general problem for them. According to Spirtes, Glymour and Scheines (p. 44), a causal arrow gets filled in on a graph between two random

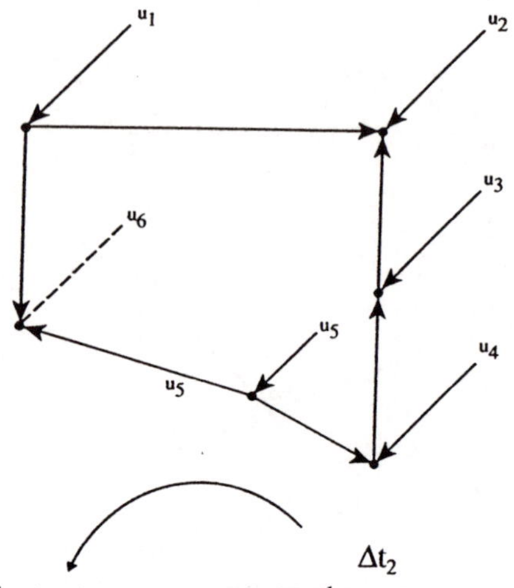

In this structure, w_1 causes w_2 and w_5 causes w_4 and w_6.

Figure 5.7b. Source: Towfic Shomar.

Returning to the main argument, I wanted this foray into Simpson's paradox to provide a concrete example of the harmful methodological consequences of a mistaken elevation of the description of the supporting nomological machine into the set of causes and effects under study. The work on causal structures by Spirtes, Glymour and Scheines is very beautiful: they provide powerful methods for causal inference, and their methods are well suited to a number of real situations in which the methods work very well. But if we model the situations in the wrong way – with the causal structure in the vertex set – the theorems will in general no longer apply. In such circumstances no reason can be provided why these methods should be adopted. The right way to think about them, I urge, is with nomological machines. The causal structure arises from a nomological machine and holds only conditional on the proper running of the machine; and the methods for

variables if any value of the first causes any value of the second. So anytime that causal influences are of different strengths for different values, we have one graph representing two different situations. There are even cases where for one pair of values, one variable causes another, and for a different pair, the second causes the first. In this case, under their prescription the graph will not even be acyclic.

Second, they say they are interested in qualitative relations – whether or not X causes Y – rather than in the quantitative strength of the causal relation, and that they do not like results that depend on particular values for those strengths. Nevertheless the plausibility of their Markov condition presupposes that they are studying sets of causal relations that are stable not only qualitatively but also across changes in the quantitative strengths of influence as well.

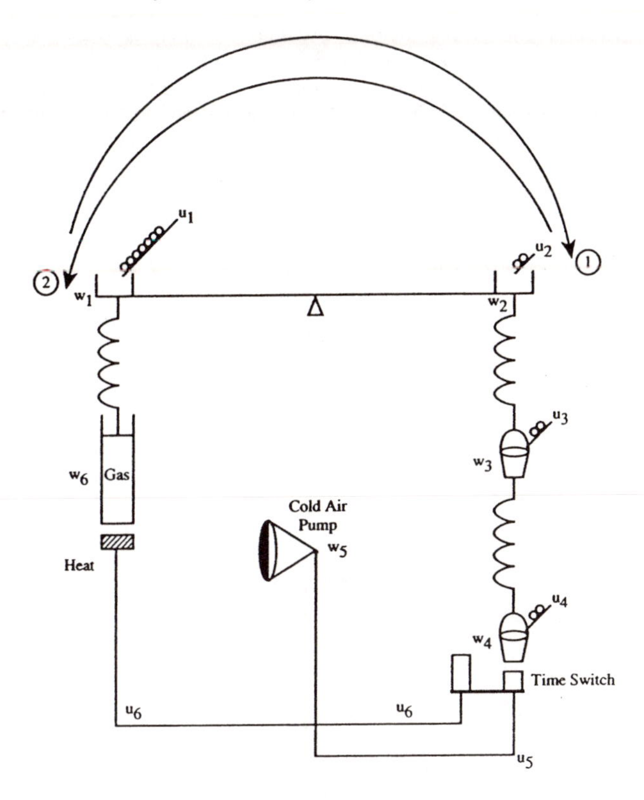

Figure 5.8 The nomological machine giving rise to the structures of figure 5.7. Source: Towfic Shomar.

studying nomological machines are different from those we use to study the structures they give rise to. Unfortunately these methods do not yet have the kind of careful articulation and defence that Spirtes, Glymour and Scheines and the Pearl group have developed for treating causal structures.

ACKNOWLEDGEMENTS

This chapter is dedicated to Elizabeth Anscombe, from whom I learned, 'The word "cause" itself is highly general ... I mean: the word "cause" can be *added* to a language in which are already represented many causal concepts' (Sosa and Tooley (eds.) 1993, p. 93; reprinted from Anscombe 1971), and also that for many causes, 'shields can be put up against them' (*ibid.*, p. 100).

Sections 1 to 2.5 of this chapter are drawn from material to be published in Cartwright forthcoming (b). The bulk of section 3 is taken from Cartwright 1997a, though new arguments have been added, as well as the discussion of the Faithfulness condition.

Thanks to Jordi Cat for very helpful discussions as well as to the participants of the British Society for the Philosophy of Science Conferences in 1991 and 1993 and the participants of the Notre Dame Causality in Crisis Conference. Thanks also to Towfic Shomar for comments, graphs and help in the production. Research for this chapter was supported by the LSE Research Initiative Fund and the LSE Modelling and Measurement in Physics and Economics Project.

6 *Ceteris paribus* laws and socio-economic machines

1 Why *ceteris paribus* laws are excluded from economics

Economics, we are told, studies laws that hold only *ceteris paribus*.[1] Does this point to a deficiency in the level of accomplishment of economics; does it mean that the claims of economics cannot be real laws? The conventional regularity account of laws answers 'yes'. On this account a theoretical law is a statement of some kind of regular association,[2] either probabilistic or deterministic, that is usually supposed to hold 'by necessity'. The idea of necessity is notoriously problematic. Within the kind of empiricist philosophy that motivates the regularity account it is difficult to explain what constitutes the difference between law-like regularities and those that hold only by accident, 'nonsense' correlations that cannot be relied on. I shall not be primarily concerned with necessity here; I want to focus on the associations themselves.

As I have rehearsed in chapter 3, empiricism puts severe restrictions on the kinds of properties that appear in Nature's laws, or at least on the kinds of properties that can be referred to in the law-statements we write down in our theories. These must be *observable* or *measurable* or *occurrent*. Economists are primarily concerned with what is measurable, so that is what I shall concentrate on here. It also restricts the kinds of facts we can learn. The only claims about these quantities that are admissible into the domain of science are facts about patterns of their co-occurrence. Hence the specification of either an equation (in the case of determinism) or of a probability distribution over a set of measurable quantities becomes the paradigm for a law of nature. These two assumptions work together to ban *ceteris paribus* laws from the nature that we study. Together they tell us that all the quantities we study are qualitatively alike. There are no differences between them not fixed either by the patterns of their associations or by what is observable or measurable about them. As a consequence it becomes impossible to find any appropriate

[1] *Cf.* Hausman 1992, ch. 8, Blaug 1992, pp. 59–62, Hutchison 1938, pp. 40–46.
[2] Hempel and Oppenheim 1960.

category that would single out conditioning factors for laws from the factors that fall within them.

What then could it mean to include as a law in one's theory that (to use an example I will discuss later) 'In conditions *C*, all profitable projects will be carried out'? All that is law-like on the Humean picture are associations between measurable quantities. That's it. The only way a condition could restrict the range of an association in a principled or nomological way would be via a more complex law involving a general association among all the quantities in question. The effect of this is to move the conditioning factor *C* inside the scope of the law: 'All projects that are both profitable and satisfy conditions *C* will be carried out.' That is what the laws of nature that we are aiming for are like. If theories in economics regularly leave out some factors like *C* (whatever *C* stands for) from the antecedents of their laws, and hence write down law claims that turn out to be, read literally, false, then economics is regularly getting it wrong. These theories need to keep working on their law-statements till they get ones that express true regularities.

As we have seen, I defend a very different understanding of the concept of natural law in modern science from the 'laws = universal regularities' account I have been describing. We aim in science to discover the natures of things; we try to find out what capacities they have and in what circumstances and in what ways these capacities can be harnessed to produce predictable behaviours. The same is true in theoretical economics.[3] Regularities are secondary. Fixed patterns of association among measurable quantities are a consequence of the repeated operation of factors that have stable capacities arranged in the 'right' way in the 'right kind' of stable environment: regularities are a consequence of the repeated successful running of a socio-economic machine. The alternative account of laws as regularities goes naturally with a covering law theory of prediction and explanation. One set of regularities – the more concrete or pheomenological – is explained by deducing them from another set of regularities – the more general and fundamental. The distinction is like that in econometrics between structural equations (like the demand and supply equations of chapter 3) and reduced-form equations (like the simple equations that tell us what the single allowed value of the price is). As I urged in chapter 4, the alternative theory of explanation in terms of natures rejects the covering law account. You can not have regularities 'all the way down'.

Ceteris paribus conditions play a special role when explanation depends

[3] I use 'theoretical' in this case to contrast the kind of model I look at in this chapter with models that we arrive at primarily by induction, for instance, large macro-economics econometric models.

on natures and not on covering laws. On the natures picture there is a general and principled distinction between the descriptions that belong inside a law statement and those that should remain outside as a condition for the regularity described in the law to obtain. The regularities to be explained hold only *ceteris paribus*; they hold relative to the implementation and operation of a machine of an appropriate kind to give rise to them. The hypothesis I defend in this chapter is that the covering-law account is inappropriate for much of the work in modern economic theory.

The detailed example I will present is one in which we cross levels of scale, or aggregation, in moving from *explanandum* to *explanans*. The model I will describe is one in which a (highly idealised) macroeconomic regularity is shown to arise from the rational behaviour of individual agents. This crossing of levels is not important to the point. We could attempt to explain macroeconomic regularities by reference to capacities and relations that can only be attributed to institutions or to the economy as a whole, with no promise of reduction to features of individuals.[4] On the other hand we could undertake – as psychologists often do – to explain behavioural or psychological regularities by reference to behavioural or psychological characteristics of the individuals involved. The point is that in all these cases it is rarely laws, in the sense of regular associations among observable or measurable quantities, that are fundamental. Few, if any, economic regularities, no matter how basic or important or central or universal we take them to be, can stand on their own. If there are laws in the regularity sense to record in economic theories, that is because they have been created.

Economists usually talk in terms of models rather than theories. In part they do so to suggest a more tentative attitude to the explanations in question than would be warranted in a fully-fledged well-confirmed theory; and in part they do so to mark a difference in the degree of articulation: theory is a large-scale (and not necessarily formalised) outline whereas a model gives a more specified (and formalised) depiction. But I think the terminology also marks a difference of importance to the hypothesis of this chapter. 'Theory' suggests a set of covering laws like Maxwell's equations, laws of the kind physics explanations are supposed to begin from. Models in economics do not usually begin from a set of fundamental regularities from which some further regularity to be explained can be deduced as a special case. Rather they are more appropriately represented as a design for a socio-economic machine which, if implemented, should give rise to the behaviour to be explained. I illustrate this idea in Section 3 with a game theory example.

[4] For an account of how to do non-reductionistic social science, see Ruben 1985 or Phillips 1987.

2 What is wrong with regularity accounts of natural law?

The most popular regularity account has it that *laws of nature* are *necessary regular associations between GOOD (= sensory or measurable or occurrent) properties*. What is wrong with this account? The answer is – everything. There are four ingredients in the characterisation, and each fails. My major concern in this chapter is with regularities in economics, so I shall discuss them at length. Before that, though, for the sake of completeness, let me review what goes wrong with each of the others.

Necessity. The usual objection is to the introduction of modalities. I argue by contrast that modal notions cannot be eliminated. The problem with the usual account is that it uses the wrong modality. We need instead a notion akin to that of *objective possibility*. I have discussed this briefly in section 6, chapter 3, and will say no more about it here because this issue, though important, is tangential to my central theses about the dappled world.

Association. Causal laws are central to our stock of scientific knowledge, and contrary to the hopes of present-day Humeans, causal laws can not be reduced to claims about association. In fact if we want to know what causal laws say, the most trouble-free reconstruction takes them to be claims about what kinds of singular causings can be relied on to happen in a given situation. For instance, 'Aspirins relieve headaches in population *p*' becomes 'In *p*, it can be relied on that aspirins do cause headache relief some of the time.' I do not of course wish to argue that facts about associations do not matter in science; nor to deny that we can construct machines like those in chapter 5 to produce causal laws. Rather I wish to point out that laws must not be understood in terms of association since we have both lawful associations and causal laws, and neither is more fundamental than the other. Again since this issue is not central to the concerns of this book, I shall not go into any more detail here.[5]

Sensible/measurable/occurrent properties. Economists like to stick with properties that are easy to observe and to which we can readily assign numbers. I have no quarrel with measurable properties – when they are indeed properties. But we should remember that we can, after all, produce a lot of nonsense measurements. Consider, for example, David Lewis' nice case in which we give the graduate record examination in philosophy to the students in his class, but we mark the results with the scoring grid for the French examination. Our procedure satisfies all the requirements of the operationalist – it is public, well articulated and gives rise to results that command full intersubjective agreement. But the numbers do not represent any real property of the students. Real properties bring specific capacities with them,

[5] For an extended defence of these claims, see Cartwright 1989.

like the capacity Lewis' students should have to tell a *modus ponens* argument from a *modus tollens* or the capacity to look 'red' to 'normal' observers or to enrage a bull. As I have argued in chapter 3, once we have given up Hume's associationist account of concept formation, I do not think we can find anything wrong with the very idea of a capacity, nor, as I argued in Cartwright 1989, is there anything epistemically second-rate about capacities. Nevertheless, in certain specific scientific contexts we may wish to restrict ourselves to making claims about readily measurable quantities or to using only 'occurrent property' descriptions for a variety of reasons. We may, for example, wish to ensure a uniform standard across different studies or different contexts. But we will not thereby succeed in describing a nature stripped of capacities, powers and causes. Nor will we succeed in producing true law claims purely in a vocabulary like this, since the concealed *ceteris paribus* conditions will always have modal force. This point is defended at length in a number of chapters in this book, so I need not go into it further here. Instead I turn to *regularity* which I have so far not treated with sufficient attention.

Regularity. The most immediate problem with regularities is that, as John Stuart Mill observed,[6] they are few and far between. That is why economics cannot be an inductive science. What happens in the economy is a consequence of a mix of factors with different tendencies operating in a particular environment. The mix is continually changing; so too is the background environment. Little is in place long enough for a regular pattern of associations to emerge that we could use as a basis for induction. Even if the situation were stable enough for a regularity to emerge, finding it out and recording it would be of limited use. This is the point of Trygve Haavelmo's example of the relation between the height of the throttle and the speed of the automobile.[7] This is a useful relation to know if we want to make the car go faster. But if we want to build a better car, we should aim instead at understanding the capacities of the engine's components and how they will behave in various different arrangements.

The second problem with regularities is that, as in physics, most of the ones there are do not reflect the kind of fundamental knowledge we want, and indeed sometimes have. We want, as Mill and Haavelmo point out, to understand the functioning of certain basic rearrangeable components.[8] Most of what happens in the economy is a consequence of the interaction of large numbers of factors. Even if the arrangement should be appropriate and last

[6] Mill 1836, see also Mill 1872.

[7] Haavelmo 1944.

[8] As I use the term here, features are 'fundamental' to the extent that their capacities stay relatively fixed over a wide (or wide enough) range of circumstances.

long enough for a regular association to arise (that is, they make up a socio-economic machine) that association would not immediately teach us what we want to know about how the parts function separately. Hence the laments about the near impossibility in economics of doing controlled experiments designed especially to do this job. What we need to know is about the capacities of the distinct parts. We may ask, though, 'Are not the laws for describing the capacities themselves just further regularity statements?' But we know from chapters 3 and 4 that the answer to this is 'no'.

Let us turn next to the idea of a mechanism *operating on its own*, say the supply mechanism or the demand mechanism. We may conceive of the demand mechanism in terms of individual preferences, goals and constraints or alternatively we may conceive it as irreducibly institutional or structural. In either case on the regularity account of laws the law of demand records the regular behaviour that results when the demand mechanism is set running alone. This paradigmatic case (which Haavelmo himself uses in talking about laws for fundamental mechanisms) shows up the absurdity of trying to describe the capacities of mechanisms in terms of regularities. No behaviour results from either the supply or the demand mechanism operating on its own, and that is nothing special about this case. In general it will not make sense to talk about a mechanism operating on its own. That is because in this respect economic mechanisms really are like machine parts – they need to be assembled and set running before any behaviour at all results. This is true in even the most stripped-down cases. Recall the lever. A rigid rod must not only be affixed to a fulcrum before it becomes the simple machine called a lever; it must also be set into a stable environment in which it is not jiggled about.

The same is true in economics. Consider an analogous case, a kind of economic lever that multiplies money just as the lever multiplies force. In this example banking behaviour will play the role of the rigid rod with 'high-powered money' as the force on one end. The reserve ratio will correspond to where the fulcrum is located. Here is how the money-multiplying mechanism works. The 'central bank' has a monopoly on making money. Following convention, let us call this 'high-powered money'. The commercial banking system blows up, or multiplies, the high-powered money into the total money supply. The banks do this by lending a fraction of the money deposited with them. The larger the proportion they can lend (*i.e.*, the smaller the reserve ratio) the more they can expand money.

Two factors are at work: the proportion of high-powered money held as currency, which is one minus the proportion deposited; and the proportion lent, which is one minus the reserve ratio. Suppose, for example, that high-powered money = £100. All of it could be deposited in a bank. That bank could lend £80. All £80 could be deposited in another bank. That bank could lend £64, and so on. The total of all deposits, $100 + 80 + 64 + \ldots = 500$. Assume that banks lend all they can, which as profit makers they are disposed to do. Then we can derive

$$M = \frac{1 + cu}{re + cu} H \tag{1}$$

where H = high powered money; re = the reserve ratio; cu = the rate of currency to deposits; M = the money stock. Equation (1) is like the law of the lever. Regularity theorists would like to read it as a statement of regular association *simpliciter*. But that is too quick. We do not have a description of some law-like association that regularly occurs. Rather we have a socio-economic machine that would give rise to a regular association if it were set running repeatedly and nothing else relevant to the size of the money stock happens.[9]

This simple example illustrates both of the central points I want to argue in this chapter. First, regularity theorists have the story just upside down. They mistake a sometime-consequence for the source. In certain very felicitous circumstances a regularity may occur, but the occurrence – or even the possibility of the occurrence – of these regularities is not what is nomologically basic. We want to learn how to construct the felicitous circumstances where output is predictable. Our theories must tell us what the fundamental mechanisms available to us are, how they function, and how to construct a machine with them that can predictably give rise to a regular association. And the information that does this for us is not itself a report of any regular association – neither a real regular association (one that occurs) nor a counterfactual association (one that might occur). Second, the example shows the special place of *ceteris paribus* conditions. Equation (1) holds *ceteris paribus*, that is under very special circumstances we may designate C. C does not mark yet another variable like H, M, cu and re that has mistakenly been omitted from equation (1), either by ignorance or sloth. The *ceteris paribus* clause marks the whole description of the socio-economic machine that if properly shielded and run repeatedly will produce a regular rise and fall of money with deposits of the kind characterised in equation (1).

I use the word 'mechanism' in describing the money multiplier. Regularity theorists also talk about mechanisms. But for them a mechanism can only be another regularity, albeit a more 'fundamental' one. 'More fundamental' here usually means either more encompassing, as in the relation of Newton's to Kepler's laws, or having to do with smaller parts as in the relation between individuals with their separate behaviours and expectations on the one hand

[9] My thanks to Max Steuer for help with this example.

and macroeconomic regularities on the other. Economics does not have any fundamental laws in the first sense. But the second sense will not help the regularity theorist either since explanations citing regularities about individuals have all the same problems as any other. When we say 'mechanism' I think we mean 'mechanism' in the literal sense – 'the structure or adaptation of parts of a machine'.[10] The little banking model is a good case. In our banking example the fixed relationship between high-powered money that the central bank creates and the ultimate size of the money stock depends on the currency-to-deposit ratio not changing and on every commercial bank being fully lent up, that is, being tight on, but not beyond, the legal reserve ratio. It is not immediately obvious that a group of commercial banks can 'multiply' the amount of currency issued by the central bank when no one bank can lend more than is deposited with it. The banking system can lend more even though no one bank can. How much more? That is what the 'law' tells us. But it is not derived by referring to a regularity, but rather by deducing the consequences of a mechanism. I illustrate the relations between economic models and regularities with another example in section 3.

When confronted with the fact that there seem to be a lot more laws than there are regularities, regularity theorists are apt to defend their view by resorting to counterfactual regularities. Laws are identified not with regularities that obtain but with regularities that would obtain if the circumstances were different. It is fairly standard by now to go on to analyse counterfactuals by introducing an array of other 'possible' worlds. Laws then turn out to be regularities that occur 'elsewhere' even though they do not occur here – that is, even though they do not occur at all. As a view about what constitutes a law of nature this 'modalised' version of the regularity account seems even more implausible than the original. A law of nature of this world – that is, a law of nature *full stop*, since this is the only world there is – consists in a regularity that obtains nowhere at all.

There are two strategies, both I think due to David Lewis, that offer the beginnings of an account of how possible regularities can both be taken seriously as genuine regularities and also be seen to bear on what happens in the actual world. The first takes possible worlds not as fictional constructions but as real.[11] The second takes them as bookkeeping devices for encoding very complicated patterns of occurrences in the real world. In this case the truth of a counterfactual will be entirely fixed by the complicated patterns of events that actually occur. Rather than employing this very complicated semantics directly we instead devise a set of rules for constructing a kind of chart from which, with the aid of a second set of rules, we can read off

[10] *The Concise Oxford Dictionary* (eighth edition).
[11] Lewis 1986.

whether a counterfactual is true or false. The chart has the form of a description of a set of possible worlds with a similarity relation defined on them. But the description need not be of anything like a world, the 'worlds' need not be 'possible' in any familiar sense, and the special relation on them can be anything at all so long as a recipe is provided for how to go from facts about our world to the ordering on the possible worlds. The trick in both cases, whether the possible worlds are real or only function as account books, is to ensure that we have good reasons for making the inferences we need; that is, that we are able to infer from the truth of a counterfactual as thus determined to the conclusion that if we really were repeatedly to instantiate the antecedent of the counterfactual, the consequent would regularly follow. This of course is immediately guaranteed by the interpretation that reads claims about counterfactual regularities just as claims about what regularities would occur if the requisite antecedents were to obtain. But then we are back to the original question. What sense does it make to claim that laws consist in these nowhere existent regularities?

The point of this question is to challenge the regularity theorist to explain what advantage these non-existent regularities have over a more natural ontology of natures which talks about causes, preventations, contributing factors, triggering factors, retardants and the like. We may grant an empiricist point of view in so far as we require that the claims of our theories be testable. But that will not help the regularity theorist since causal claims and ascriptions of capacities or natures are no harder to test than claims about counterfactual regularities and indeed, I would argue, in many cases you can not do one without the other.[12]

3 An example of a socio-economic machine

The game-theoretic model proposed by Oliver Hart and John Moore in their 'Theory of Debt Based on the Inalienability of Human Capital' provides a good example of a blueprint for a nomological machine.[13] I pick this example not because it is especially representative of recent work in economic theory but rather because the analogy with machine design is transparent in this case and the contrast with a covering law account is easy to see. The central idea behind the model is that crucial aspects of debt contracts are determined by the fact that entrepreneurs cannot be locked into contracts but may withdraw with only small (in their model *no*) penalties other than loss of the project's assets. That means that some debt contracts may be unenforceable and hence

[12] I have defended the second of these claims in Cartwright 1989 and in chapter 4. The first is argued for in chapter 3.

[13] Hart and Moore 1991; republished with minor changes as Hart and Moore 1994.

inefficiency may result, *i.e.*, some profitable projects may not be undertaken. Hart and Moore derive a number of results. I shall discuss only the very first (corollary 1) as an illustration of how socio-economic machines give rise to regularities. As Hart and Moore describe, 'Corollary 1 tells us that inefficiency arises only if either (a) there is an initial sunk cost of investment . . . and/or (b) the project's initial returns are smaller than the returns from the assets' alternative use . . .'.[14]

The model presents a toy machine that if set running repeatedly generates an economically interesting regularity described in corollary 1. The inputs can vary across the identity of individual players, sunk costs, income streams, liquidation-value streams, and the initial wealth of the debtor. The output we are considering is the regularity described in corollary 1:

R: All profitable projects which have no initial sunk costs and whose initial returns are at least as large as the returns from the alternative use of the assets will be undertaken.

The model lays out a number of features necessary for the design of a machine: It tells us (i) the parts that make up the machine, their properties and their separate capacities, (ii) how the parts are to be assembled and (iii) the rules for calculating what should result from their joint operation once assembled. The components are two game players ('debtor' or 'entrepreneur', and 'creditor') who (a) have the same discount rates, (b) are motivated only by greed, (c) operate under perfect certainty and (d) are perfect and costless calculators. The arrangements of the players is claustrophobic: two players set against each other with no possible interaction outside. Later extensions consider what happens when either the debtor or the creditor have profitable reinvestment opportunities elsewhere, but these are fixed opportunities not involving negotiations with new players. Another extension tries in a very 'indirect and rudimentary fashion'[15] to mimic with just the two players what would happen if the debtor could negotiate with a new creditor. Other assumptions about the arrangement are described as well; for instance, as part of the proof of R it is assumed 'for simplicity' that the debtor 'has all the bargaining power' in the original negotiation.[16] The central features of the arrangement are given by the rules laid out for the renegotiation game, plus a number of further 'simplifying' assumptions about the relations among capital costs, project returns and liquidation returns that contribute to the proof of R. As we saw in chapter 3, rules for calculating how parts function

[14] Hart and Moore 1991, p. 19. The result need mention only initial returns because of the simplifying assumption Hart and Moore make that if the return at any period is greater then the liquidation income at that period, this will continue to be the case in all subsequent periods.

[15] *Ibid.*, p. 40.

[16] *Ibid.*, p. 17.

together vary widely across domains.[17] In game theory various concepts of equilibria tell us how to calculate the outcomes of the joint action of players with specified capacities (*e.g.*, perfectly or imperfectly informed, costless calculators, etc.). For the Hart and Moore model it is supposed that the moves of the players together constitute a subgame perfect equilibrium.

The model is less informative about the two remaining features necessary for describing the operation of a machine: (iv) what counts as shielding and (v) how the machine is set running. As we have seen, Hart and Moore describe a very closed world. There are just two players and they are locked into the very special game they are playing. The shielding conditions are implicit. Nothing must occur that distorts the rules of the game and nothing must affect the choice of debt contracts other than the requirement that they constitute a subgame perfect equilibrium. Repeated running simply means playing the game again and again. This is consistent with my suggestion in chapter 3 that models like this are described at a very high level of abstraction. The model specifies abstract functional relations between the parts that can be instantiated in various different institutional arrangements; what counts as shielding will depend heavily on what the specific material instantiation is. This is especially true in game-theoretical models, where few clues are given about what real institutional arrangements can be taken to constitute any specific game.

To derive the regularity R, Hart and Moore show that in their model all subgame perfect equilibria for their date zero renegotiation game satisfy a set of conditions (proposition 1) that in turn imply corollary 1. The derivation proceeds by unpacking the concept of equilibrium in an exogeneously given extensive form game. In particular the derivation employs no general laws that might be mistaken for claims about regularities. There are no empirical generalisations nor any equations on the analogue of Schrödinger's or Maxwell's written down by Hart and Moore; nor any general theorems of game theory. Rather the argument employs concepts and techniques of game theory (some of which are of course validated by theorems) plus the general characterisation of equilibrium to derive constraints on what the pay-offs will be from any equilibrium in the game described.

Turn now to the questions about *ceteris paribus* conditions and the regularity theory of laws. R holds (at best) only *ceteris paribus*: if conditions like those of the model were to occur repeatedly, then in those situations all profitable ventures with no sunk costs and good enough initial returns would be undertaken. On the picture of natural law that I have been advocating, something like the converse is true as well. If a regularity like R is to obtain

[17] This is part of the reason it is difficult to formulate a general informative philosophical account of mechanisms and their operation.

as a matter of law, there must be a machine like the one modelled by Hart and Moore (or some other, with an appropriate structure) to give rise to it. There are no law-like regularities without a machine to generate them. Thus *ceteris paribus* conditions have a very special role to play in economic laws like R. They describe the structure of the machine that makes the laws true.

The relation of laws to models I describe is familiar in economics, where a central part of the theoretical enterprise consists in devising models in which socio-economic regularities can be derived. But it is important to realise how different this is from the regularity theory. Look back to the regularity theory. R is not a universal association that can be relied on outside various special arrangements. On the regularity theory law-likeness consists in true universality. So there must be some universal association in the offing or else R cannot be relied on at all, even in these special circumstances. If an association like R appears to hold in some data set, that cannot be a matter of law but must be viewed as merely a chance accident of a small sample unless there is some kind of true universal association to back it up.

The difference between an account of natural law in terms of nature's capacities and machines and the regularity account is no mere matter of metaphysics. It matters to method as well, both the methods used in the construction of theory and those used in its testing. R tells us that *ceteris paribus* all profitable ventures will be taken up – except in conditions (a) and (b). The regularity theory invites us to eliminate the *ceteris paribus* clause by extending this list to include further factors – (c), (d), (e), . . ., (x) – until finally a true universal is achieved: 'All profitable ventures that satisfy (a), (b), . . ., (x) will be taken up.' This way of looking at it points the investigation in an entirely wrong direction. It focuses the study on more and more factors like (a) and (b) themselves rather than on the structural features and arrangements like those modelled by Hart and Moore that we need to put in place if we want to ensure efficiency.

The regularity theory also carries with it an entourage of methods for testing that have no place here. When is a model like that of Hart and Moore a good one? There are a large number of different kinds of problems involved. Some are due to the fact that theirs is a game-theoretic model; these are to some extent independent of the issue raised by the differences between a regularity and a capacity view of laws. The advantage to game theory is that it makes the relationship between the assumptions of the explanatory model and laws like R that are to be explained in the model very tight. The results that are 'derived' in the model are literally deduced. The cost is that the rules of the games that allow these strict deductions may seem to be very unrealistic as representations of real life situations in which the derived regularities occur. As Hart and Moore say about their own model, 'The game may seem *ad hoc*, but we believe that almost any extensive form bargaining game is subject to this criticism.'[18] This is an example of the kind of process I described in the Introduction by which economics becomes exact – but at the cost of becoming exceedingly narrow. The kind of precise conclusions that are so highly valued in contemporary economics can be rigorously derived only when very special assumptions are made. But the very special assumptions do not fit very much of the economy around us.

Let us lay aside these special problems involving games and think about the relations between models and the results derived within them more generally. My point is similar to one stressed in chapter 5: the results derived may be rendered as regularity claims, but the relationship between the structures described by the model and the regularities it gives rise to is not again itself one of regular association. So our whole package of sophisticated techniques – mostly statistical – for testing regularity claims are of no help in the decisions about choice among models. How do we decide? As far as I can see we have no articulated methodology, neither among philosophers nor among economists (though we may well have a variety of unarticulated methods). My reasons for attending to the metaphysics of economic laws stem primarily from these methodological concerns. So long as we are in the grip of the regularity view – which happens to economists as well as to philosophers – we are likely to restrict our efforts at methodology to refining ever more precisely our techniques at statistics and to leave unconsidered and unimproved our methods for model choice.

4 Three aspects of socio-economic machines

'Socio-economic laws are created by socio-economic machines.' The slogan points to three distinct theses, which are separable and can be argued independently. The first, the one that I am most prepared to defend, follows Aristotle in seeing natures as primary and behaviours, even very regular behaviours, as derivative. Regular behaviour derives from the repeated triggering of determinate systems whose natures stay fixed long enough to manifest themselves in the resulting regularity. This feature does not point particularly to a machine analogy, though, in opposition to an organic one, as is apparent from the work of Aristotle himself.

Organic analogies usually suggest a kind of irreducible wholism. The behaviours of the components when separated have little to teach about how they work together in the functioning organism. By contrast, the machine analogy stresses the analytic nature of much economic thought. This is the second thesis: much of economic work is based on the hope that we can understand different aspects of the economy separately and then piece the lessons

[18] *Ibid.*, p. 12.

together at a second stage to give a more complete account. This idea is central to the use of idealisation and correction that is common throughout economics. Bounded rationality is expected to be like unbounded but with modifications; international trade at best tends to move prices towards each other but they are often modelled as being equal; in planning models, the planner is often assumed to have no other goal than social welfare just as the firm manager is assumed to maximise share-holder value of the firm; and so forth. I do not want to urge this second thesis suggested by talking of socio-economic machines as vigorously as the first. There is no guarantee that the analytic method is the right method for all the problems that economics wants to treat. But it is, I think, by far and away the one we best know how to use.

The third thesis is one about which evidence is divided. Ordinary machines do not evolve. They have to be assembled and the assembly has to be carefully engineered. Is the same true of socio-economic machines? I do not intend in using the description to urge that the answer must be 'yes'. One of the most clear-cut examples[19] of a 'designed' institution in economics is the International Monetary Fund (IMF), which resulted from negotiations over two radically different designs put forward for an international body to run post-Second-World-War international monetary arrangements. In large part the design adopted at Bretton Woods in 1944 was for a 'fund' from which member countries could draw previously deposited reserves, rather than a 'bank' which could create credit on deposits. Economic historians have tended to rate the institution a failure because of its design faults: (1) it was not devised to deal with the transition from the war years to the peace years; (2) it ran out of reserves (as predicted by Keynes) in the 1960s and so had to create SDR (special drawing rights) to service the growth in the international economy; and (3) it was unable to bring pressure to bear on surplus countries to revalue, compromising its ability to avoid damaging currency crises and oversee orderly changes in exchange rates, leading ultimately to the collapse of the arrangements in 1971. By contrast economic historians have tended to rate the gold standard of the pre-1914 economy as a much more effective institution – an institution which was not designed but which evolved gradually over the nineteenth century. They suggest that this institution worked well because its evolutionary character allowed it to be operated with considerable flexibility and tacit knowledge, rather than because it stuck to some supposed 'rules of the game' or was operated according to some agreed economic theory.[20]

[19] I owe the following discussion to Mary Morgan.
[20] A standard good source is Foreman-Peck 1983.

5 Economics versus physics

I began with the conventional claim that the laws of economics hold only *ceteris paribus*. This is supposed to contrast with the laws of physics. On the regularity account of laws this can only reflect an epistemological difference between the two. Economists simply do not know enough to fill in their law claims sufficiently. I have proposed that, if there were a difference, it would have to be a metaphysical difference as well. Laws in the conventional regularity sense are secondary in economics. They must be constructed, and the knowledge that aids in this construction is not itself again a report of some actual or possible regularities. It is rather knowledge about the capacities of institutions and individuals and what these capacities can do if assembled and regulated in appropriate ways.

Does this really constitute a difference between economics on one hand and physics on the other? As we have seen in earlier chapters, I think not. It is sometimes argued that the general theory of relativity functions like a true covering-law theory. It begins with regularities that are both genuinely universal and true (or, 'true enough'); the phenomena to be explained are just special cases of these very general regularities. Perhaps. But most of physics works differently. Like economics, physics uses the analytic method. We come to understand the operation of the parts – for example, Coulomb's force, the force of gravity, weak and strong nuclear interactions, or the behaviour of resistors, inductors and capacitors – and we piece them together to predict the behaviour of the whole. Even physics, I argue, needs 'machines' to generate regularities – machines in the sense of stable configurations of components with determinate capacities properly shielded and repeatedly set running. If this is correct then differences in the metaphysics of natural laws that I have been describing are not differences between economics and physics but rather between domains where the covering-law model obtains and those where the analytic method prevails. Economics and physics equally employ *ceteris paribus* laws, and that is a matter of the systems they study, not a deficiency in what they have to say about them.

ACKNOWLEDGEMENTS

The bulk of this chapter is taken from Cartwright 1995a. The review of problems with the regularity account of laws in section 2 is new. Research for this paper was sponsored by the Modelling and Measurement in Physics and Economics Project at the Centre for Philosophy of Natural and Social Science, London School of Economics. I would like to thank the members of that research group for help with this paper, as well as especially Mary Morgan and Max Steuer. The paper was read at a symposium on realism in the Tinbergen Institute; I would also like to thank the students and faculty who participated in that symposium for their helpful comments.

7 Probability machines: chance set-ups and economic models

1 Where probabilities come from

Ian Hacking in *The Logic of Statistical Inference*[1] taught that probabilities are characterised relative to chance set-ups and do not make sense without them. My discussion is an elaboration of his claim. A chance set-up is a nomological machine for probabilistic laws, and our description of it is a model that works in the same way as a model for deterministic laws (like the Newtonian model of the planetary system that gives rise to Kepler's laws discussed in chapter 3 or the socio-economic machine of chapter 6). A situation must be like the model both positively and negatively – it must have all the relevant characteristics featured in the model and it must have no significant interventions to prevent it operating as envisaged – before we can expect repeated trials to give rise to events appropriately described by the corresponding probability.

2 Who is the enemy?

The view I am arguing against takes probabilities to be necessary associations between measurable quantities; and it assumes that probabilities in this sense hold not *ceteris paribus* but *simpliciter*. In particular they do not need a nomological machine to generate them. I am going to discuss here a case from economics in which it is easy to see the implausibility of a free-standing probability. It is typically objected that physics is different. We have seen in the previous chapters why I do not agree. For here let me just recall how odd the concepts of physics are for the kind of empiricism that still expects scientific concepts to come with some test procedures attached. Contrary to strict empiricist demands, the concepts of physics generally do not pick out independently identifiable or measurable properties. Its concepts, rather, build in – and thus cover up – the whole apparatus necessary to get a nomological machine to work properly: the arrangement of parts, the shielding and what-

[1] Hacking 1965.

ever it takes to set the machine running. Remember the case of the lever from chapter 3. A lever is not just a rigid rod appropriately affixed to a fulcrum. For we do not call something a 'lever' unless the rod is *appropriately affixed and shielded* so that it will obey the law of the lever. From the point of view of empiricism, physics regularly cheats. Economists tend to be truer to empiricism. In formulating laws they do try to stick to properties that are measurable in some reasonable sense. This is one of the reasons why I want to illustrate my views about machines with socio-economic examples.

The example I will consider in this section involves a debate about Sri Lanka and the case for direct government action to maintain the food entitlement of the population. I focus on a discussion fairly late in the debate, by Sudhir Anand and Ravi Kanbur.[2] Sri Lanka has been considered as a test case in development strategy because of its 'exceptionally high achievements in the areas of health and education'.[3] For a long while Sri Lanka has had a very high life expectancy and high educational levels in comparison with other developing countries in spite of being one of the low-income group. Anand and Kanbur state the point at issue thus: 'The remarkable record in achievement is attributed by some to a systematic and sustained policy of government intervention in the areas of health, education, and food over a long period. The counter to this position comes in several forms, which can perhaps best be summarised in the statement that the intervention was, or has become, "excessive" relative to the achievement.'[4] Anand and Kanbur criticise a 'cross-sectional' model standard in the literature, for which they want to substitute a time series model. Their criticisms are detailed and internal.[5] What I want to object to is the whole idea of using a cross-sectional model of this kind to think about the success of Sri Lanka's welfare programme.

Here is their model, which is intended 'to explain some measure of living standard, H_{it}, for country i at time t':[6]

$$H_{it} = \alpha_t + \beta Y_{it} + \delta E_{it} + \lambda_i + u_{it}$$

where Y_{it} is per capita income; E_{it} is social welfare expenditure; α_t is a time-specific but country-invariant effect assumed to reflect technological advances; λ_i is a country-specific and time-invariant 'fixed effect'; δ is the marginal impact of social expenditure on living standards; and u_{it} is a random error term.

The trick is to estimate δ: the marginal impact of social expenditure on

[2] Anand and Kanbur 1995.
[3] *Ibid.*, p. 298.
[4] *Ibid.*
[5] And not, to my mind, correct. *Cf.* Sen 1988.
[6] Anand and Kanbur 1995, p. 321.

living standards. This is supposed to exist and be fixed. It represents the strength of the effect that direct government expenditure will exert on standard of living in a developing country. That is what I find so surprising, to think that there could be any such relationship just given for developing countries, as though laid down by the hand of God in the Book of Nature. It seems to me that this methodology is just crazy. That is because this equation represents a 'free-standing' probability: there is no good reason to think there is a chance set-up that generates it. You can of course use statistics as what the word originally meant – a *summary* of some data. In so far as there are data about the relevant quantities in a number of countries, you can write it out in lists or you can summarise it in statistics. And you can do that whether or not there is any probability over the quantities at all since probability is irrelevant to the use of statistics as summaries. The use intended for δ, though, requires a real probability: a law-like regular association. Why should we think that there is any such law-like association? To suppose that there really is some fixed probability relation between welfare expenditure and welfare like that in the equation you need a lot of good arguments. You need arguments both at the phenomenological level – the frequencies pass strong tests for stability – and at the theoretical level – you have got some understanding of what kind of shared socio-economic structure is at work to generate them. So far my review of the literature has not turned up very good arguments of either kind.

The equation I have written down comes from the work by Anand and Kanbur criticising Amartya Sen for estimating the wrong equation: for estimating an equation in the level of the quantities rather than in their first differences. But this is odd since Sen does neither. Rather than the single graph pictured in figure 7.1, which expresses a hypothesis Anand and Kanbur criticise about the causal relations that hold among the designated quantities across all developing countries, Sen's hypotheses dictate different graphs for different countries. And he does it this way exactly for the reasons I say. Each of the countries studied has a different socio-economic structure constituting a different socio-economic machine that will generate different causal relations true in that country and concomitantly different probability measures appropriate for the quantities appearing in these relations.

Look just at how Sen talks about South Korea and Taiwan on the one hand versus Sri Lanka on the other.[7] His causal hypotheses for these two countries are pictured in figures 7.2a and 7.2b, which should be contrasted with the causal structure of figure 7.1. In Taiwan and South Korea the immediate level of causation for poverty removal on his account is employment expansion with growing demand for labour, and in this case export expansion was the

[7] Sen 1981.

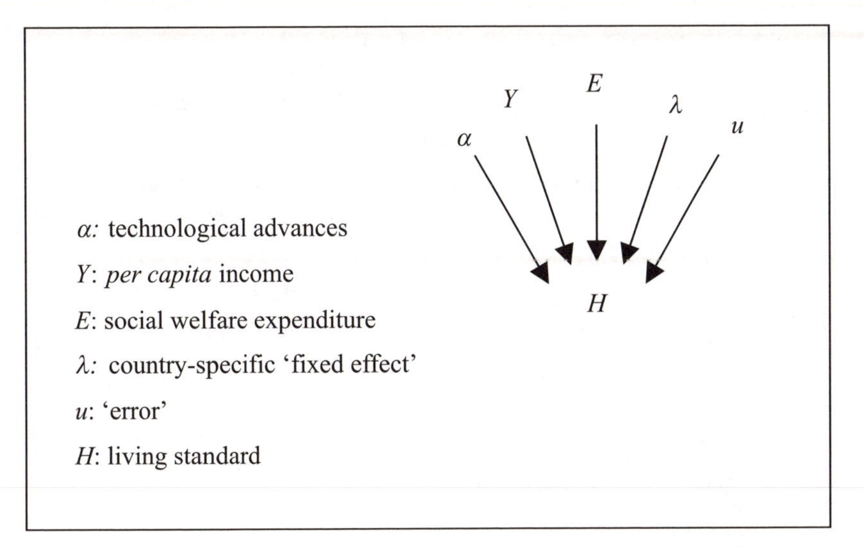

Figure 7.1 Anand-Kanbur's 1995 causal model for standard of living.

immediate cause of the growing demand for labour. But, considering South Korea, 'the fact that export expansion provided a very good vehicle for such explanation . . . has to be viewed in the perspective of the total economic picture.'[8] Sen tells us '*behind* [my emphasis] the immediate explanation'[9] stand lots of other factors, like rapid capital accumulation, availability of a labour force suited by education to modern manufacture, and 'overwhelming control of the organised banking sector'[10] by the government. When these kinds of factors are arranged appropriately they constitute a kind of socio-economic machine in which causal links between export expansion and poverty removal can arise.

Before turning to possible replies to my attack on free-standing laws, I should note that I am making a point similar to one popular in economics in the last twenty years, though not new to them, since it was at the heart of John Stuart Mill's argument that economics could not be an inductive science. These are points often expressed in the vocabulary of autonomy and stability. Macroeconomic models like the standard-of-living equation are not likely to be stable, since equations like this depend on some further underlying structure remaining fixed. This is a favourite line of the Chicago School to undermine calls for government action. But they add to this two-tiered

[8] *Ibid.*, p. 298.
[9] *Ibid.*, p. 299.
[10] *Ibid.*, p. 298.

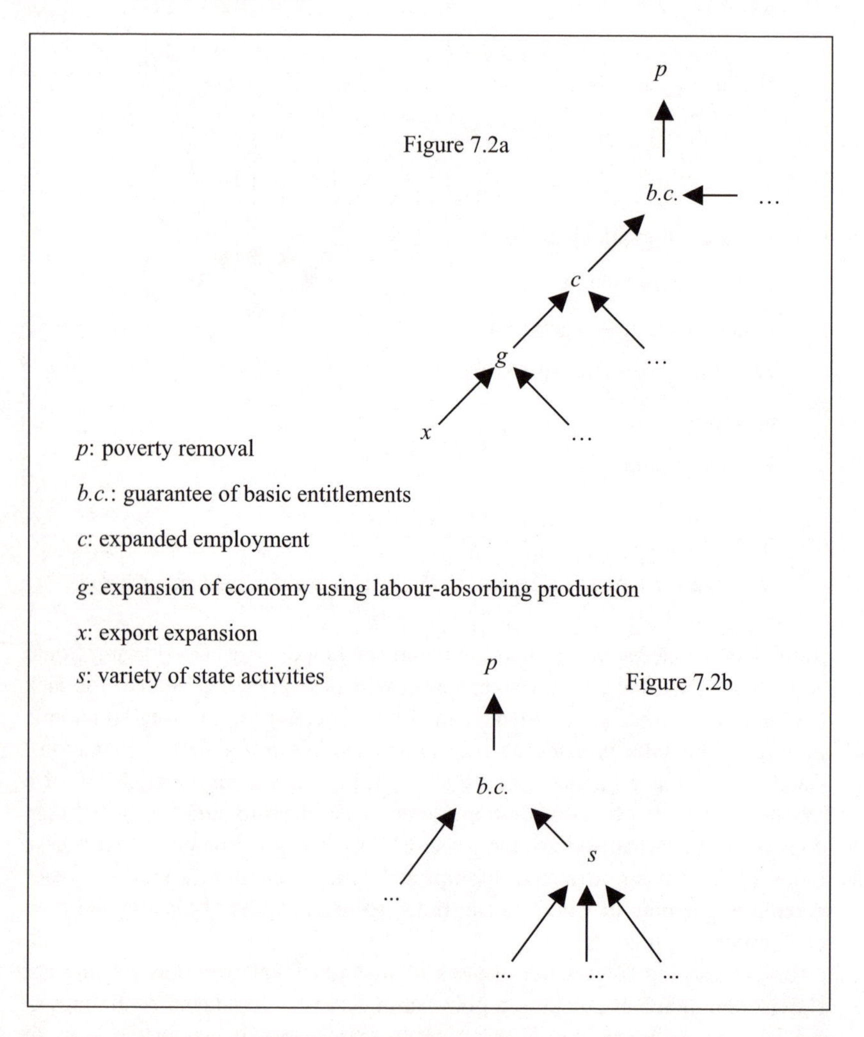

Figure 7.2: Sen's causal structure for (a) South Korea and Taiwan and (b) Sri Lanka.

picture I defend two further assumptions to get their conclusion. (1) The underlying structure must be one of individual rational decisions. (2) The way these decisions are made will always after the short term readjust to undo whatever a government intervention might try to achieve. I note these two additional assumptions since there is no reason in the arguments I give to think that we need have a micro base in order to account for macro regularity. The planetary system is a nomological machine: the capacities of the planetary masses to attract one another when appropriately ordered and unhindered generate Kepler's laws. There is no reason that the components of the machine and their capacities need be individuals rather than institutions. And nothing I say here supports the specialist assumptions that the Chicago School use to generate their conclusions about what happens when agents plan ahead for what the government might do.

3 Where probabilities do not apply

What do social scientists say about my worries about using free-standing probabilities like those in the standard-of-living/welfare expenditure study? Econometrician David Hendry assumes that econometrics would ideally like to discover 'the complete structure of the process that generates economic data and the values of all its parameters'.[11] He seems to presuppose that this process is always appropriately modelled with a probability measure. In conversation he has defended this assumption with the remark that the sampling procedure can ensure that a probability measure is appropriate. I can make sense of this from the point of view of the social statistician looking at samples from a particular population in order to infer characteristics of the population as a whole.[12] In this case it makes a lot of difference how one thinks the sample has been drawn.

Consider a simple example. My population is ten playing cards: four hearts, two diamonds, four spades. You are to draw one card at random. What is the conditional probability that the card drawn is a diamond, given it is red? Here there is an answer: $\frac{1}{3}$. Next consider a more complicated procedure. First we do two independent flips of a fair coin. If the results are HH, you draw one card at random from the 'pointed' cards in my population (*i.e.*, diamonds and spades), otherwise from the non-pointed cards. Again, it is entirely correct to assign a probability measure to the outcomes, and under that measure the conditional probability that a drawn card is a diamond given it is red is $\frac{1}{4}$.

But now let us imagine I give the cards to the persons sitting closest to me to order on their aesthetic preferences; then I pick the top card. Now we have to argue the case, and I do not know any good reasons to suggest that this situation should be represented by any one or another of the models of chance set-ups to which probability measures attach. What argument is there to show that a data-generating method like this, characterised by features outside the domain of probability theory, must have an accurate (or accurate

[11] Hendry 1995, p. 7.

[12] This understanding of Hendry is supported by his own example of the dice rolling experiment: Hendry 1995, p. 8.

enough) description to bring it into the theory? On the other hand, to insist that every data-generating process, regardless of its features, is correctly described by some probability measure or other is just to assert the point at issue, not to defend it.

We can of course construct an analogous situation for the developing countries, where the conditional probability for the standard of living given a certain level of welfare expenditure does make sense. Imagine you represent a medical insurance company in a position to set up business in one of the developing countries. Which of these countries you will be allowed to invest in will be assigned in a fair lottery. You will not be able to find out the standard of living in the country you draw but information on its level of welfare expenditure will be available. In this case I have no quarrel with the existence of a conditional probability for a country you might draw to have a given standard of living for a given level of expenditure – that is just like the probability that a card drawn from my pack will be a diamond given it is red; and indeed it will be extremely useful to you in representing the insurance company to know what this conditional probability is.

I have been talking throughout about laws of nature in the realist way that is typical nowadays, and I have been assuming the usual connection between laws and warranted prediction. Laws I claim, including probabilistic laws, obtain in only very special kinds of circumstances. Concomitantly, rational prediction is equally limited. For great swathes of things we might like to know about, there are no good grounds for forecasting. Glenn Shafer in his recent book, *The Art of Causal Conjecture*,[13] develops his own special account of probability, but he arrives at the same conclusion. Shafer's standpoint is that of a staunch empiricist. His probabilities have directly to do with rational prediction, without any mediation by laws. Shafer finds the concepts of a *law governing* the evolution of events and of the *determination* of one state of affairs by another to be problematic. I take it that his views are similar to those of our Vienna Circle forebears. Perhaps we know how to begin to think about matters concerning *our* making something happen. But what guides do we have when it comes to one natural event forging another into a prescribed form, much less that of laws of nature writ nowhere by no one constraining the behaviour of objects in their dominion?

Shafer's alternative is to take probability as the rational degree of belief of a Laplacean demon armed with complete information about everything that has happened so far in history. The demon thus has just the same *kind* of information that we do, only lots more of it. And the demon is engaged in just the *kind* of activity that we engage in daily in science and thus have pa firm grasp on. It is very important in Shafer's thinking that the Laplacean

[13] Shafer, 1996.

demon – whom he calls 'Nature' – has the same kind of information that we do.[14] This means for him, first, that Nature knows no laws but only facts; second, that the facts, even when they are all known up to a certain time, may exhibit no coherent pattern. They may continue to look just as disorderly as they do to real practising scientists, who know only a small subset of them. This I take it is why Shafer argues that for many cases, probability – *i.e.*, Nature's rational degrees of belief – may not exist. Consider the diagram in figure 7.3a. Shafer says:

> Consider the determination of a woman's level of education in the society represented [on the left hand side of figure 7.3a]. Perhaps Nature can say something about this. Perhaps when Nature observes certain events in a woman's childhood she changes her predictions about how much schooling the woman will receive. [The right hand side of figure 7.3a] gives an example, in which we suppose that the experience of being a girl scout encourages further schooling. But it is not guaranteed that Nature will observe such regularities. Perhaps the proportion of girls becoming scouts and the schooling received by scouts and non-scouts varies so unpredictably that Nature cannot make probabilistic predictions, as indicated in [figure 7.3b]. Perhaps there are no signals that can help Nature predict in advance the amount of schooling a girl will get.[15]

Figure 7.3b is especially interesting because probability gaps appear in the middle of a tree. I too have been arguing for probability gaps like these. Recall the example of chapter 5. I sign a cheque for £300, as I did this past weekend to pay our nanny. I suppose that the banking system in England for ordinary transactions is well regulated, the rules are clearly laid out, and the bank employees are strongly disposed and competent to follow the rules. They do have a policy reviewing the signatures on cheques, however, which leads to a small number of false rejections. Coupling that with other failures (like an inattentive clerk) we assume that not all properly signed cheques for £300 cause the person to whom they are written – in this case, our nanny – to be £300 richer if they reach that person's bank. Still there is a very good chance they will: we may say 'a high qualitative probability'. But that qualitative probability is not free standing. Rather it obtains on account of the institutional set-ups that I described – crudely, the banking laws and people's dispositions to obey them.

Notice, however, that I said the payee should be made richer by the cheque signing *if* the cheque arrives. That is in general not a problem. There are a thousand and one ways, or more, to get a cheque from here to there, and all guaranteed by some different institutional arrangements than those

[14] *Ibid.*
[15] Shafer 1997, pp. 7–8.

that make possible the causal law connecting cheque signing with the payee's getting money. In fact it makes sense to set up the banking machine in the way we do just because there are a variety of ways readily available to carry the causal message from the cause to the effect. They make it possible for there to be a high probability for the payee to become richer, given that a cheque is signed; but there is no necessity that they themselves have any analogous conditional probability. A signed cheque must travel by bus or mail or train or foot or messenger or some other way in order to be successful at providing money to the payee. But that goes no way to showing that there is some conditional probability or other that a cheque goes by, say, Royal Mail, given it is signed.

Think about my situation for example. The children were staying with me in London and would see the nanny on the way back to boarding school. I was flying to California the next day. Would I send the cheque by hand with them, or post it with the Royal Mail, or, too busy to buy English stamps, take it all the way to California and use the United States mail? Nothing, I think, fixes even a single case propensity, let alone a population probability. Certainly if there is one, it is not guaranteed by the same nomological machine that fixes the possibility for a conditional probability for a payee to receive money given that a cheque is signed, nor by the many institutions that guarantee the existence of the various routes the cheque may follow: for example, the charter for the Royal Mail that runs the post; the Department of Transport that keeps up the roads; the statutes and attitudes that prohibit apartheid so that my children will be able to enter Oxford; and so forth. In order to ensure that there is a conditional probability that a cheque is, say, in the mail given it is signed, one would need another nomological machine beyond all these, one that structured the dispositions of the banking population in a reliable way. I do not think there is one. But that of course is an empirical question. The point is that these new conditional probabilities are not required by the other probabilistic and causal laws we have assumed we have;[16] and if they are to obtain we will need some different socio-economic machines beyond the ones we have so far admitted.

The mention of conditional probabilities should remind us of another range of cases where we have long recognised that there may be probability gaps: cases where the conventional characterisation of conditional probability – $\mathrm{Prob}(B/A) = \mathrm{Prob}(A\&B)/\mathrm{Prob}(A)$ – seems to fail. For instance situations where the conditional probability seems easy to assess, but the quantities in

[16] Shafer 1996 uses the theory of martingales to establish a number of results about when 'new' probabilities, either qualitative or quantitative, will be required by probabilistic assumptions already made.

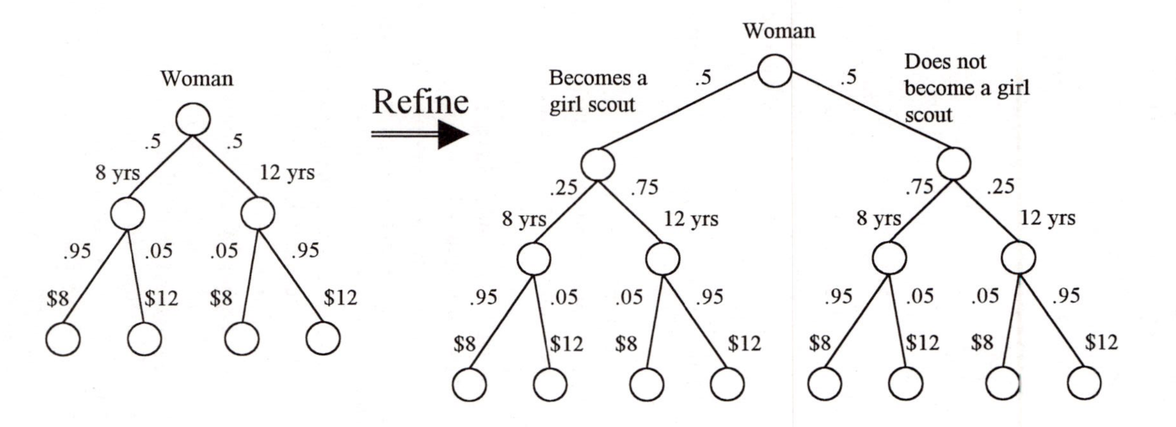

Some detail about how the educational level of a woman is determined. Notice that the refinement agrees with all the causal assertions in the original tree. When a woman leaves school after eight years, nature gives her a 5% chance of earning $12. How she decided to leave school does not matter.

Figure 7.3a. Source: Shafer 1997, p. 8.

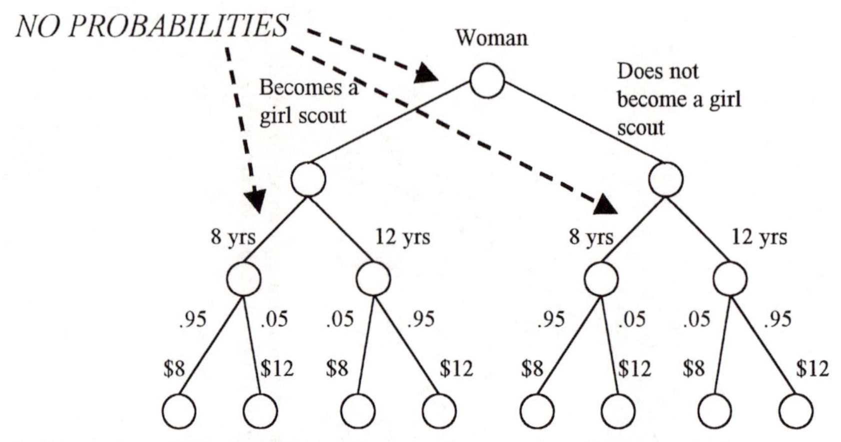

In this version of the story, Nature does not observe any stable pattern in the proportion of girls who become girl scouts or in the proportion of girl scouts and non-girl scouts who finish 12 years of schooling.

Figure 7.3b Source: Shafer 1997, p. 8.

the definition do not make sense. Consider an example of Alan Hájek's.[17] What is the probability that a fair coin lands heads, given that I toss it? Hájek takes the answer to be obvious: one half, of course. Yet the probability that I toss the coin is undefined, he urges.

This is just what I would say as well. But unlike Hájek I do not think that this particular kind of example shows up faults with our definition of conditional probability. For this case is not like my card example where we calculate the conditional probability that we have drawn a diamond given that the card we have drawn is red. That is, it is not a case in which we are asked to consider the probability of certain outcomes obtaining conditional on certain other *outcomes* occurring. For my tossing the coin is not an outcome at all: there is no chance set-up that produces it. Instead we have here a paradigm case of a nomological machine (the tossing of a fair coin) giving rise to a probability distribution (the probability of landing heads or tails).

A similar example arises in discussions of the Bell inequalities in quantum mechanics. Here a pair of particles is supposed to be in a specific quantum state, the *singlet* state. If we make separate measurements on the spins of the two particles at quite distant locations, we find that the outcomes are correlated. This is surprising because it seems that neither particle has a spin beforehand but only 'decides' at the point of measurement what outcome will

[17] Hájek 1997b.

be displayed; nor, given the set-up, does it seem possible for the particles to communicate with each other at the last moment about what the outcome will be. How then do the results come to be correlated? One hypothesis is that the initial quantum state – the singlet state – acts as a joint cause producing the two outcomes in tandem. How should we test that hypothesis? One natural idea is to make use of the standard probabilistic analysis of causality. According to this analysis, (*ceteris paribus*) C causes E if $\text{Prob}(E/C) > \text{Prob}(E/-C)$. So for each of the particles we consider the question: is the probability of a spin-up outcome higher given the singlet state than it is without the singlet state?

Now we are back at Hájek's problem with the coin tossing example. If we have a well constructed experimental arrangement with particles prepared and maintained in the singlet state, quantum mechanics can tell us the probability of a given outcome in a spin measurement. But if we try to read that as a conditional probability, as the probabilistic analysis of causation requires, we are in trouble. Corresponding to $\text{Prob}(C\&E)$ in this case is the probability that we create the right kind of experimental arrangement involving a singlet state and then get a spin-up outcome. Quantum theory has nothing to say about what this probability could be; and it does not look as if any other theory will do the job either. The point is even more striking when we consider $\text{Prob}(E/-C)$, for now we have to calculate what the probability of a spin-up outcome is if we measure a particle *not* prepared in the singlet state. What state, if any, should the particle be in then, and what should be the probability of this state? We do not have an answer. Nor, I think, does anyone expect one. We intuitively recognise that here there is no probability to be had.

4 Two examples of probability-generating machines

I begin with an example familiar to philosophers of science: Wesley Salmon's argument that causes can decrease as well as increase the probability of their effects.[18] Salmon considered two causes of a given effect, one highly effective, the other much less so. When the highly effective cause is present, he imagined, the less effective one is absent, and vice versa. Thus the probability of the effect goes down when the less effective cause is introduced even though it is undoubtedly the cause of the effect in question. That is the story in outline, but exactly what must be the case to guarantee that the probabilities are (a) well defined and (b) that they fall within appropriate ranges to ensure the desired inequality [Prob(effect/less effective cause) < Prob(effect/ – less effective cause)]?

[18] Salmon 1971.

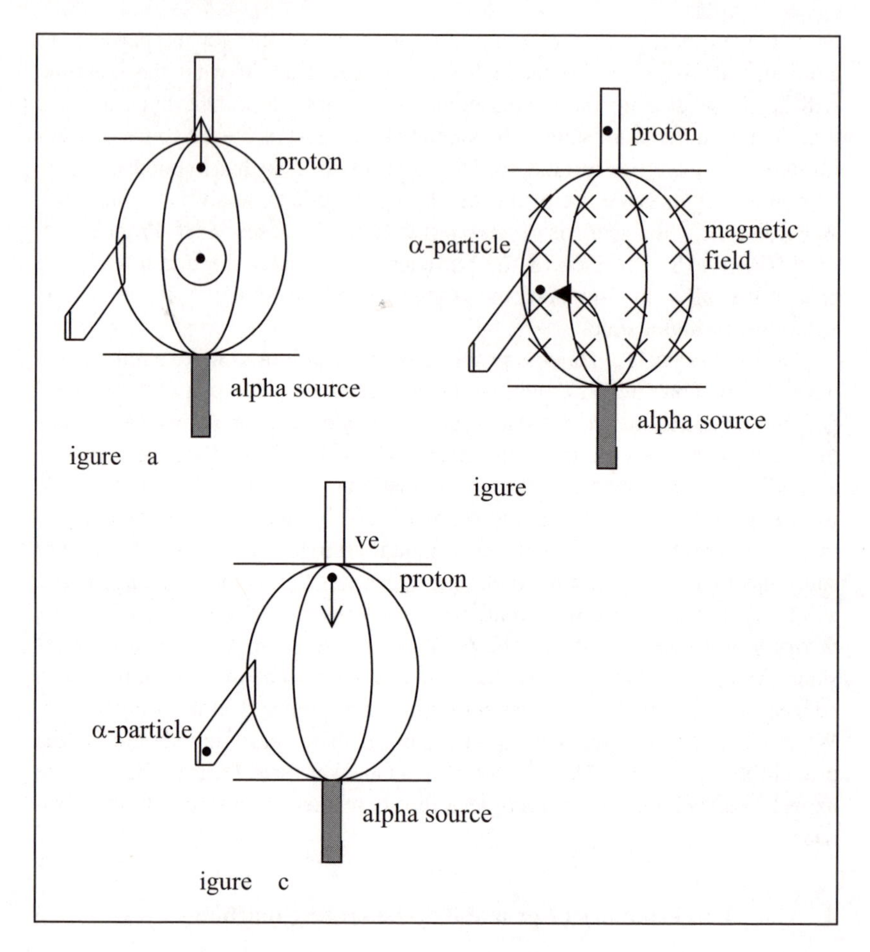

Figure 7.4. Salmon machine. Source: designed by Towfic Shomar

First we need an arena – say, a closed container. Then we need a mechanism, a mechanism that can ensure two things: that there is a fixed joint probability (or range of fixed probabilities with a fixed probability for mixing) for the presence and absence of the two kinds of causes in the container, and that under this probability there is sufficient anticorrelation, given the levels of effectiveness, to guarantee the decrease in probability. Next we must ensure that there is no other source of the effect in the container or introduced with either of the elements. We must also guarantee that there is nothing present in correlation with either of the causes that annihilates the effect as it is produced, etc., etc. Figure 7.4 is a model of the kind of arrangement that

A radioactive source that emits an alpha particle on average every 15 minutes is installed in a cylindrical container opened to a spherical chamber containing a proton. The expulsion forces from an emitted alpha particle push the proton into the cylindrical box (Figure 7.4a). If the alpha particle is influenced by a magnetic field, it will travel to exit C (Figure 7.4b). A magnetic field is turned on in the chamber for 15 minutes and is cut off for 15 minutes; when it goes off the upper cylinder will become positively charged, forcing the proton back into the chamber (Figure 7.4c). (We can assume that the positive charge at the upper cylinder influences the system for no more than half a minute, allowing the proton to enter.) Let $c =_{df}$ the presence of an alpha particle, $e =_{df}$ the presence of the proton in the top cylinder and $m =_{df}$ the presence of a magnetic field in the chamber. Then we have

Prob(e/c) > Prob($e/\text{-}c$) Prob(e/c) is high (0.9)
Prob($c\&m$) = 0 Prob(e/m) is low (0.1)

Since Prob($e/\text{-}m$) = Prob(e/c), we can conclude that Prob(e/m) < Prob($e/\text{-}m$) even though m causes e.

Figure 7.4 cont.

is required – a model for a chance set-up or a probability-generating machine. Salmon himself used radioactive materials as causes, the effect being the presence or absence of a decay product.

My experiment is designed by Towfic Shomar, who chose different causes to make the design simple. The point is that without an arrangement like the one modelled (or some other very specific arrangement of appropriate design) there are no probabilities to be had; with a sloppy design, or no design at all, Salmon's claims cannot be exemplified.

As a second example, consider David Hendry and Mary Morgan's discussion of the views of Trygve Haavelmo. Haavelmo is one of the founders of econometrics, who, years after the fact, won a Nobel prize for his important work developing the probability approach in economics. Haavelmo worried about what is still a problem in econometrics: the difference between parameters and variables. His account of the difference is difficult to understand. But the solution to this problem is not our concern here. Rather I introduce the issue so that we can see how Hendry and Morgan deploy probability generating machines in their efforts to make sense of Haavelmo's solution. They say:

The problem at this stage for Haavelmo is that all his concepts are interrelated, and we offer the following example to try and clarify the issues. Imagine tossing a perfectly balanced coin. By symmetry, the probability of a head equals that of a tail and is a half (excluding the low probability of it landing on its edge). The design of experiments in mind is that the tosses are independent and undertaken with a randomly varying force, and we construe the probability of a half as the parameter and the outcomes as the random variable. Now imagine the tossing being done by a perfectly calibrated machine in a vacuum, where the coin is always loaded with the same face up in the machine. If the outcome on the first toss is a head, the probability of a head becomes unity in successive tosses. Here, changing the design of experiments has radically altered the parameter and the realizations. Consequently, all aspects of the design of experiments must be selected jointly.[19]

In Hendry and Morgan's example we have one coin with a fixed physical structure, but two probability machines. We could of course have a vast variety of other machines generating a vast variety of other probabilistic behaviours. We may if we like say of the symmetric coin that it has a fifty per cent probability to land heads and fifty per cent to land tails. But that is just a shorthand way to refer to a very generic capacity the coin has by virtue of being symmetric. In this case we are picking out the capacity by implicitly pointing to the behaviour it gives rise to in what we all recognise to be a canonical chance set-up. But as Hendry and Morgan make clear, with different surrounding structures the coin will give rise to different probabilities.

Imagine by contrast that we flip the coin a number of times and record the outcomes, but that the situation of each flip is arbitrary. In this case we cannot expect any probability at all to emerge.[20] Just notice how specific Hendry and Morgan are in their description of the two machines. In the first, 'the tosses are independent and undertaken with a randomly varying force'; in the second, we 'imagine the tossing being done by a perfectly calibrated machine in a vacuum, where the coin is always loaded with the same face up in the machine'; and clearly in both cases they mention only a few of the necessary features they are assuming the set-ups to have. The matter is different of course if by 'arbitrary' tosses we mean 'random' tosses, either independent or with some special kind of fixed dependencies, shielded so that nothing further affects the outcome than what has already been captured by previous descriptions. In this case we have a probability. But we also have a well understood chance set-up. With no fixed surrounding structure of the right kind for the flips, there will be no probabilities at all to describe the outcomes.

5 How probability theory attaches to the world

Turning from these specific examples, we can ask, 'In general, how do probabilities attach to the world?' The answer is *via models*, just like the abstract descriptions of physics that I discussed in chapter 2. Just like the classical

[19] Hendry and Morgan 1995, p. 68.
[20] Nor a mix of probabilities.

Ch. 6 Probability Distributions
6.1 Introduction
6.2 The Hypergeometric Probability Distributions
6.3 The Binomial Probability Distributions
6.4 The Poisson Probability Distributions
Parameters of the Poisson
Use of the Poisson
Expected Gain of the Poisson Probability Distributions
6.5 The Normal Probability Distributions
Converting other Normal Distributions to the Statistical Normal Distribution
Applications of Converted Normal Distributions
6.6 The Exponential Poisson Distributions
6.7 Approximation with Substitute Distributions
The Binomial as an Approximation to the Hypergeometric
The Poisson as an Approximation to the Hypergeometric and the Binomial
The Normal Curve Approximation to the Hypergeometric and the Binomial
An Overall Comparison of Approximation Results

Figure 7.5 Probability distributions. Source: McAllister 1975.

force function, the electromagnetic field vectors and the quantum Hamiltonian, a probability measure provides an abstract description of a situation, a description that always piggy-backs on a specific set of more concrete descriptions. So assigning a probability to a situation is like assigning a force function or a Hamiltonian: set distributions are associated with set descriptions. The distributions listed in the table of contents of Harry E. McAllister's *Elements of Business and Economic Statistics*,[21] shown in figure 7.5, are typical. (Compare, for instance, Kyburg 1969, Mulholland and Jones 1968 or Robinson 1985.) Further familiar distributions appear later: the t-distribution, the Chi-square distribution and the F-distribution.

As in physics, where the description of a situation that appears in the model must be specially prepared before the situation can be fitted to the theory (*e.g.*, once you call something a 'harmonic oscillator', then mechanics can get a grip on it), so too in probability theory. The description of events as *independent* and as *equally likely* or of samples as *random* are key, as we have seen in the example from Hendry and Morgan. We can illustrate with the simple binomial distribution, which McAllister describes this way: 'A

[21] McAllister 1975.

large class of problems applicable to situations in business and economics deals with events that can be classified according to two possible outcomes ... If, in addition to having two possible outcomes, the outcome of one particular trial is independent of other trials and if the probability ... is constant for other trials, the binomial probability distribution can be applied.'[22]

Again as in physics, in learning probability theory we are taught a handful of typical or paradigmatic examples to which, *ceteris paribus*, the prepared descriptions of the model may be applied (*e.g.*, vibrating strings, pendula and the modes of electromagnetic fields may all be treated as harmonic oscillators). So probability theory too has its stock examples that show what kinds of concrete material descriptions are likely to support the theoretical descriptions that must be satisfied before the theory can apply. 'Uses of the Poisson distribution', McAllister informs us, '... include the broad area of theory involving random arrivals such as customers at a drive-in bank, customers at a check-out counter and telephone calls coming into a particular switchboard'.[23] Mulholland and Jones repeat the example of telephone calls in a given period, adding particle emission and the number of minute particles in one millilitre of fluid, as well as a caution: 'But there must be a random distribution. If the objects have a tendency to cluster, *e.g.*, larvae eggs, then the Poisson distribution is not applicable.'[24]

The hypergeometric distribution tends to have three kinds of illustrations: defective items (especially in industrial quality control); cards (especially bridge hands) and fish sampling (without replacement of course). And so forth. In each case a given distribution will apply only to situations that have certain very specific – and, generally, highly theoretical – features. Because the requisite features are so theoretical, it is best to supply whatever advice possible about what kinds of situations are likely to possess these features. But these are just rough indications and it is the features themselves that matter: situations that have them – and have no further features that foil them – should give rise to the corresponding probability; and again as in our earlier examples, without them we get no probabilities at all. Probabilities attach to the world via models, models that serve as blueprints for a chance set-up – *i.e.*, for a probability-generating machine.

6 An economics example

So far we have looked at the chance set-ups with well-known arrangements of characteristics that feature in probability theory and the corresponding

[22] *Ibid.*, p. 111.
[23] *Ibid.*, p. 120.
[24] Mulholland and Jones 1968, p. 167.

distributions that they give rise to. I would like now to look at an example from an empirical science, in particular economics. Most economic models are geared to produce totally regular behaviour, represented, on the standard account, by deterministic laws. My example here is of a model designed to guarantee that a probabilistic law obtains.

The paper I will discuss is titled 'Loss of Skill during Unemployment and the Persistence of Unemployment Shocks' by Christopher Pissarides.[25] I choose it because, out of a series of employment search models in which thc number of jobs available depends on workers' skills and search intensities, Pissarides' is the first to derive results of the kind I shall describe about the probabilities of unemployment in a simple way. The idea investigated in the paper is that loss of skill during unemployment leads to less job creation by employers which leads to continuing high levels of unemployment. The method is to produce a model in which

f_t = the probability of a worker getting a job at period t

(i) depends on the probability of getting a job at the previous period (f_{t-1}) if there is skill loss during unemployment – i.e., shows *persistence*; and (ii) does not depend on f_{t-1} if not. The model supposes that there is such a probability and puts a number of constraints on it in order to derive a further constraint on its dynamics:

(i) $\partial f_t / \partial f_{t-1} > 0$, given skill loss

(ii) $\partial f_t / \partial f_{t-1} = 0$, given no loss of skill

The point for us is to notice how finely tuned the details of the model plus the initial constraints on the probability must be in order to fix even a well-defined constraint on the *dynamics* of f_t, let alone f_t itself.

The model is for two overlapping generations each in the job market for two periods only: workers come in generations, and jobs are available for one period only so that at the end of each period every worker is, at least for the moment, unemployed. 'Short-term unemployed' refers to 'young' workers just entering the job market at a given time with skills acquired through training plus those employed, and thus practising their skills, in the previous period; 'long-term unemployed' refers to those from the older generation who were not employed in the previous period. The probability f_t of a worker getting a job in the between-period search depends critically on χ, the number of times a job and worker meet and are matched so that a hire would take place if the job and the worker were both available. By assumption, χ at t is a determinate function of the number of jobs available at t (J_t) and the number of workers available at t ($2L$). Wages in the model are deter-

[25] Pissarides 1992.

mined by a static Nash bargain, which in the situation dictates that the worker and employer share the output equally and guarantees that all matches of available workers and jobs lead to hiring. The central features of the first model are listed in figure 7.6. Variations on the basic model that relax the assumptions that all workers search in the same way and thus have the same probability for a job match are developed in later sections of the paper.

The details of the argument that matter to us can be summarised in three steps. (I follow Pissarides' numbering of formulae, but use primes on a number to indicate formulae not in the text but that follow in logical sequence the numbered formula.)

A firm's expected profit, π_t, from opening a job at t is

$$\pi_t = [1 + f_{t-1} + (1 - f_{t-1})y] \; (Lf_t/J_t) \qquad (4)$$

where $y = 1$ represents no skill loss, $y < 1$ the opposite. It is crucial that f_{t-1} appears in this formula. It enters because the expected profit depends on the probability of a job meeting a short- and a long-term unemployed worker, which in turn depends on the number of long-term unemployed workers available and hence on the probability of employment at t-1.

B. The number of jobs will adjust so that no firm can make a profit by opening one more, which, given that the cost of opening a job is $1/k$, leads to

$$\pi_t = 1/k \qquad (5')$$

and thus, using (4), to

$$J_t = Lk \; [1 + y + (1 - y) \, f_{t-1}] \, f_t \qquad (7)$$

C. As a part of the previous argument it also follows that $J_t \geq \chi \; (J_t, 2L)$. In addition,

$$f_t = \min \{\chi(J_t, 2L), J_t, 2L\}/2L,$$

since the number of hires cannot be greater than the number of jobs or workers available nor the number of meetings that take place. Coupling these with the assumption that the function χ is homogenous of degree 1 gives

$$f_t = \min \{\chi(J_t/2L, 1), 1\} \qquad (8)$$

When $f_t = 1$ – full employment – there are no problems. So consider

$$f_t = \chi(J_t/2L, 1) \qquad (8')$$

To do so, substitute (7) into (8′) to get

$$f_t = \chi[(k/2)\{1 + y + (1 - y)f_{t-1}\} \, f_t, \, 1] = \chi(\Phi, 1) \qquad (9')$$

letting $\Phi = (k/2)\{1 + y + (1 - y) \, f_{t-1}\} \, f_t$.

We are now in a position to draw the two sought-for conclusions, begin-

1. Discrete time.
2. Two overlapping generations.
   a. Each of fixed size, L.
   b. Each generation is in the job market exactly two periods.
3. Each job lasts one period only and must be refilled at the beginning of every period.
4. The number of jobs, J_t, available at beginning of period t is endogenous.
5. Workers in each of their two life periods are either employed or unemployed.
6. a. Output for young workers and old, previously employed workers $= 2$.
   b. Output for old, previously unemployed workers $= 2y$, $0 < y < 1$. ($y < 1$ represents skill loss during unemployment.)
7. Unemployed workers have 0 output, no utility, no income. (This is relevant to calculating wages and profits.)
8. In each period all workers and some jobs are available for matching.
9. Each job must be matched at the beginning of a period to be filled in that period.
10. In each period workers and jobs meet at most one partner.
11. The number of matches between a job and a worker is designated by χ, where
    a. χ is at least twice differentiable.
    b. First derivatives in χ are positive; second, negative.
    c. χ is homogeneous of degree 1.
    d. $\chi(0, 2L) = \chi(J_t, 0) = 0$.
    e. $\chi(J_t, 2L) = \max(J_t, 2L)$.
12. There is a probability that a worker meets a job at the beginning of t, designated by f_t.
    a. f_t does not depend on what a worker does nor on whether the worker is employed or unemployed.
    b. f_t is a function only of J_t and L.
13. There is a probability that a job meets a worker at the beginning of period t.
    a. This probability is independent of what jobs do.
    b. This probability is a function only of J_t and L.
14. The cost of opening a job and securing the output as described in 6, is equal to $1/k$ (whether the job is filled or not).
15. Wages are determined by a Nash bargain.
16. Workers and employers maximise expected utility.

Figure 7.6 Assumptions of Pissarides' 1992 model 1.

ning with the second: (ii) The case where there is no skill loss during unemployment is represented by $y = 1$. (Short- and long-term workers are equally productive. See assumption 6, figure 7.6.) Then

$$f_t = \chi(kf_t, 1)$$

from which we see that f_t does not depend on f_{t-1}. Hence with no skill loss there is no unemployment persistence in this model.
(i) When there is skill loss, $y < 1$. Differentiating (9′) with respect to f_{t-1} in this case gives

$$[1 - \{d\chi/d\Phi\}\{k/2\}\{1 + y + (1 - y)f_{t-1}\}]\ [\partial f_t/\partial f_{t-1}] = (k/2)(1 - y)f_t(d\chi/d\Phi) \quad (11)$$

Then by the homogeneity of $\chi(\Phi)$,

$$\partial f_t/\partial f_{t-1} > 0$$

'Thus', as Pissarides concludes, 'the dynamics of f_t are characterized by persistence.'[26]

The trick in the derivation is to get f_t to be a determinate function (via χ) of a product of f_t and f_{t-1}. χ itself is a function of J_t and L. In this model the product form in χ is achieved by getting J_t, which is itself determined by profits to be earned from offering a job, to depend on the product of f_t and f_{t-1}. This comes about because J_t depends on the probability of a job being filled by a short- (or long-)term worker, which in turn is equal to the probability of a worker being short- (or long-)term unemployed – into which f_{t-1} enters – times the probability of a short- or long-term worker getting a job, which is indifferently f_t for both.

The derivation of persistence, where it occurs, in the rest of the models in the paper also depends on the fact that the relevant probability analogous to f_t in model 1 is, through the matching function χ, a function of the product of $f_t f_{t-1}$. The second model looks to see what happens when the number of jobs is fixed but effort of search varies between the long- and short-term unemployed. In this case the product enters not into the constant factor J of which χ is a function, but rather into the second factor, which is not now workers available but units of search effort provided by workers seeking employment. The resulting persistence here is negative ($\partial f_t/\partial f_{t-1} < 0$) which is taken to reflect the process in which low employment increases the numbers of long-term unemployed and thereby lowers the search units supplied, which in turn raises the probability of hire per worker which leads to higher search intensity and thus to more hires.

In the third model, where jobs are again endogenous but firms expect the same profit from the short- and the long-term unemployed, the product appears in the term, S_t for search units available and through that in $J_t = kf_tS_t$.

[26] *Ibid.*, p. 1377.

Since f_t here is $\chi(J_t S_t)/S_t$ for χ homogeneous of degree 1, it disappears again. The last model, where jobs are endogenous and profits differ, is more complicated. The product appears in S_t and also in J_t, which is no longer a multiple of S_t, so the product ends up in both the numerator and denominator of χ. Thus, though the dynamics of f_t are constrained to exhibit persistence, the nature of that persistence is not determinate and could differ depending on what further conditions are added to the model to fix the characteristics of the matching function, χ, which is simply hypothesised to exist.

7 Lessons of the economics model

I repeat the lesson I wish to draw from looking at Pissarides' search model. Turn again to figure 7.6. It takes a lot of assumptions to define this model and, as we have seen, the exact arrangement matters if consequences are to be fixed about whether there is persistence in the dynamics of unemployment probability or not. Those arrangements are clearly not enough to fix the exact nature of the persistence, let alone the full probability itself. In model 1, where job openings are endogenous, the dependence of jobs on workers' histories must be engineered just so, so that J_t will be a function of the product $f_t f_{t-1}$. In model 2, where the product could not possibly enter through J_t, the facts about how workers search must be aligned just right to get the product into S_t. And so forth.

My claim is that it takes hyperfine-tuning like this to get a probability. Once we review how probabilities are associated with very special kinds of models before they are linked to the world, both in probability theory itself and in empirical theories like physics and economics, we will no longer be tempted to suppose that just any situation can be described by some probability distribution or other. It takes a very special kind of situation with the arrangements set just right – and not interfered with – before a probabilistic law can arise.

As I noted at the beginning, what is special about these situations can be pointed to by labelling them *nomological machines*: they are situations with a fixed arrangement of parts where the abstract notions of *operation, repetition*, and *interference* have concrete realisations appropriate to a particular law and where, should they operate repeatedly without interference, the outcome produced would accord with that law.

8 An objection: Alan Hájek's magic trick

Alan Hájek[27] claims to be able to get probabilities out of almost anything at all. He calls this a 'probabilistic magic trick'. If he is right, this sounds like

[27] Hájek 1997a.

a powerful objection to my conclusion that we only get probabilities from very special kinds of set-ups. Let us look at Hájek's magic trick. He says:

Here is the challenge.
1. Give me an object – any object you like.
2. Give me a number between 0 and 1 inclusive – to be thought of as any probability value you like.
3. Give me a degree of approximation – any number of decimal places of accuracy that you like.

I will use the object to generate an event whose probability is the number you chose, to the accuracy you specified. Your challenge is to stump me; my claim is that you can't.[28]

In Hájek's example, he supposes you give him a car key, choose the number $1/\sqrt{2}$, and demand accuracy to three decimal places. What does he do?

Contrary to the worry that Hájek's trick will undermine my conclusion, in fact it supports it. For what Hájek does with the key is to create a very well-understood chance set-up that everyone can recognise as appropriate for generating outcome events with probability $1/\sqrt{2}$ to the required three places of accuracy. First he looks for an asymmetry on the key, say a scratch. 'Now', he says, 'I toss the key a large number of times, and record for each toss whether the key lands 'scratch UP' or 'scratch DOWN' . . . We can regard this as a sequence of Bernoulli trials.'[29] He then groups the sequences into pairs, from which he will construct a new sequence of 'heads' and 'tails', using the following rule: (up down) becomes heads; (down up) becomes tails; the others are discarded. As a result, he tells us, 'What I have can be regarded as a *simulation of a sequence of tosses of a fair coin.*'[30] After that the job is just to use the tosses of a fair coin to generate the required probability to three decimal places, a kind of process we are familiar with from Bayesian methods for assessing degrees of belief to any required degree of accuracy. We can thus fill in for ourselves how to complete the rest of the construction.

What about these tosses of the key? The important feature is that Hájek's procedure depends on the fact that we can, as he says, regard the outcomes as a sequence of Bernoulli trials. So he needs to ensure that the tosses will be independent of each other and have the same probability from trial to trial. Like McAllister and the other authors I discussed in section 5, he knows what kind of physical arrangements generally warrant these more theoretical descriptions. For instance, he will be cautious, he says, not to let the key get scuffed in order to keep the probabilities (almost) the same, and to pick a feature of it and a way of tossing it that ensures that there will be no memory

[28] *Ibid.*, p. 1.
[29] *Ibid.*, p. 2.
[30] *Ibid.*, p. 3.

of the results from toss to toss. So what seemed at first a refutation turns into a confirmation of the points I am defending here. Hájek can get a probability from almost anything. But to do so he must construct a chance set-up. The probabilities that come out can be finely tuned by adjusting the exact construction of the chance set-up. Depending on just what the chance set-up is, the probabilities can be anything, to any degree of accuracy, or they can be nothing at all if we do not treat the key in any special way.[31]

9 Chance set-ups

I began with Ian Hacking's views from *The Logic of Statistical Inference*. There Hacking urges that propensities in chance set-ups and frequencies are obverse sides of the same phenomenon. Propensities in chance set-ups give rise to frequencies; frequencies are the expression of these propensities. This is a point of view that I have tried to defend here. A chance set-up may occur naturally or it may be artificially constructed, either deliberately or by accident. In any case probabilities are generated by chance set-ups, and their characterisation necessarily refers back to the chance set-up that gives rise to them.[32] We can make sense of the probability of drawing two red balls in a row from an urn of a certain composition with replacement; but we cannot make sense of the probability of six per cent inflation in the United Kingdom next year without an implicit reference to a specific social and institutional structure that will serve as the chance set-up that generates this probability. The originators of social statistics followed this pattern, and I think rightly so. When they talked about the 'iron law of probability' that dictated a fixed number of suicides in Paris every year or a certain rising rate of crime, this was not conceived as an association laid down by natural law as they (though not I) conceived the association between force and acceleration, but rather as an association generated by particular social and economic structures and susceptible to change by change in these structures.

The same, I claim, is true of all our laws, whether we take them to be iron – the typical attitude towards the laws of physics – or of a more flexible material, as in biology, economics or psychology. J. J. C. Smart[33] has urged that biology, economics, psychology and the like are not real sciences. That is because they do not have real laws. Their laws are *ceteris paribus* laws,

[31] I think that Hájek could agree with my insistence that the chance set-up is necessary, for it in no way counters his own interpretation of probabilities as single case propensities.

[32] Hugh Mellor also thinks that probabilities, or at least objective chances, are not defined for every arbitrary situation. For instance, an effect does not confer a chance on its causes, though the causes do fix the chance of the effect (see Mellor 1995). His arguments will help to support some of my positions as well.

[33] Smart 1963.

and a *ceteris paribus* law is no law at all. The only real laws are, presumably, down there in fundamental physics. I put an entirely different interpretation on the phenomena Smart describes. As we have seen, if the topic is *laws* in the traditional empiricist sense of claims about necessary patterns of regular association, we have *ceteris paribus* laws all the way down: laws hold only relative to the chance set-ups that generate them.

What basic science aims for, whether in physics or economics, is not primarily to discover laws but to find out what stable capacities are associated with features and structures in their domains, and how these capacities behave in complex settings. What is fundamental about fundamental physics is that it studies the capacities of fundamental particles. These have the advantage, supposedly, of being in no way dependent on the capacities of parts or materials that make them up. That is not true of the capacities of a DNA chain or a person. But the laws expressing the probabilistic regularities that arise when these particles are structured together into a nucleus or an atom are every bit as dependent on the stability of the structure and its environment as are the regularities of economics or psychology. A nucleus in an atom may be a naturally occurring chance set-up, but it is a set-up all the same. Otherwise the probabilistic laws of nuclear physics make no sense. I repeat the lesson about the dual nature of frequencies and propensities: probabilities make sense only relative to the chance set-up that generates them, and that is equally true whether the chance set-up is a radio-active nucleus or a socio-economic machine.

ACKNOWLEDGEMENTS

Sections 1, 5, 6, 7 and 9, as well as the discussion of Wesley Salmon's example in section 4 are taken from Cartwright 1997d. A portion of section 2 is from Cartwright 1997b, but the discussion of Sen's positive account is new. Sections 3 and 8 are entirely new, as is the discussion of the Hendry and Morgan example in section 4. Research for this paper was sponsored by the LSE Modelling and Measurement in Physics and Economics Project.

Part III

The boundaries of quantum and classical physics and the territories they share

8 How bridge principles set the domain of quantum theory

1 Theory and the building of models

In the 1960s when studies of theory change were in their heyday, models were no part of theory. Nor did they figure in how we represent what happens in the world. Theory represented the world. Models were there to tell us how to change theory. Their role was heuristic, whether informally, as in Mary Hesse's neutral and negative analogies, or as part of the paraphernalia of a more formally laid out research programme, as with Imré Lakatos. The 1960s were also the heyday of what Fred Suppe dubbed 'the received view' of theory,[1] the axiomatic view. Theory itself was supposed to be a formal system of internal principles on the one hand – axioms and theorems – and of bridge principles on the other, principles meant to interpret the concepts of the theory, which are only partially defined by the axioms. With the realisation that axiomatic systems expressed in some one or another formal language are too limited in their expressive power and too bound to the language in which they are formulated, models came to be central to theory; they came to constitute theory. On the semantic view of theories, theories are sets of models.[2] The sets must be precisely delimited in some way or another, but we do not need to confine ourselves to any formal language in specifying exactly what the models are that constitute the theory.

Although doctrines about the relation of models to theory changed from the 1960s to the 1990s, the dominant view of what theories do has not changed: theories represent what happens in the world. For the semantic view that means that models represent what happens. One of the working hypotheses of the LSE/Amsterdam Modelling Project has been that this view is mistaken. There are not theories, on the one hand, that represent and phenomena, on the other, that get represented (though perhaps only more or less accurately). Rather, as Margaret Morrison[3] put it in formulating the background to our project, models mediate between theory and the world. The

[1] Suppe 1977.
[2] Van Fraassen 1980, or Giere 1988.
[3] Morrison 1997.

theories I will discuss here are the highly abstract theories of contemporary physics. I want to defend Morrison's view of models not as constituting these theories but as mediating between them and the world.

Of course there are lots of different kinds of models serving lots of different purposes, from Hesse's and Lakatos' heuristics for theory change to Morrison's own models as contextual tools for explanation and prediction. In this discussion I shall focus on two of these. The first are models that we construct with the aid of theory to represent real arrangements and affairs that take place in the world – or could do so under the right circumstances. I call these *representative models*. This is a departure from the terminology I have used before. In *How the Laws of Physics Lie*,[4] I called these models *phenomenological* to stress the distance between fundamental theory and theory-motivated models that are accurate to the phenomena. But *How the Laws of Physics Lie* supposed, as does the semantic view, that the theory itself in its abstract formulation supplies us with models to represent the world. They just do not represent it all that accurately. Here I want to argue for a different kind of separation: theories in physics do not generally represent what happens in the world; only models represent in this way, and the models that do so are not already part of any theory. It is because I want to stress this conclusion that I have changed the label for these models.

Following the arguments about capacities initiated in chapter 10 of *Nature's Capacities and their Measurement*[5] and further developed here, I want to argue that the fundamental principles of theories in physics do not represent what happens; rather, the theory gives purely abstract relations between abstract concepts. For the most part, it tells us the capacities or natures of systems that fall under these concepts. As we saw in chapter 3, no specific behaviour is fixed until those systems are located in very specific kinds of situations. When we want to represent what happens in these situations we will need to go beyond theory and build a model, a *representative* model. And, as I described in chapter 3, if what happens in the situation modelled is regular and repeatable, these representative models will look very much like blueprints for nomological machines.

For a large number of our contemporary theories, such as quantum mechanics, quantum electrodynamics, classical mechanics and classical electromagnetic theory, when we wish to build a representative model in a systematic or principled way, we shall need to use a second kind of model. For all of these theories use abstract concepts, 'abstract' in the sense developed in chapter 2: concepts that need fitting out in more concrete form. The models that do this are laid out within the theory itself in its bridge principles. First

[4] Cartwright 1983.
[5] Cartwright 1989.

generation descendants of the Vienna Circle called these *interpretative* models and I shall retain the name even though it is not an entirely accurate description of the function I think they serve. The second kind of model I focus on then will be the interpretative model.

I begin from the assumption that it is the job of any good science to tell us how to predict what we can of the world as it comes and how to make the world, where we can, predictable in ways we want it to be. The first job of models I shall focus on is that of representing, representing what reliably happens and in what circumstances; and the first job of this chapter will be to distinguish theory from models of this kind. To get models that are true to what happens we must go beyond theory. This is an old thesis of mine. If we want to get things right we shall have to improve on what theories tell us, each time, at the point of application. This is true, so far as I can see, in even the most prized applications that we take to speak most strongly in theory's favour. This should not surprise us. Physics is hard. Putting it to use – even at the most abstract level of description – is a great creative achievement.

I used to argue this point by explaining how the laws of physics lie. At that time I took for granted the standard account that supposes that what a theory can do stretches exactly as far as its deductive consequences – what I here call the 'vending machine view' of theory. Since then I have spent a lot of time looking at how theories in physics, particularly quantum physics, provide help in making the world predictable, and especially at devices like lasers and SQUIDs whose construction and operation are heavily influenced by quantum mechanics. I have been impressed at the ways we can put together what we know from quantum theory with much else we know to draw conclusions that are no part of the theory in the deductive sense. The knowledge expressed in physics' fundamental principles provides a very powerful tool for building models of phenomena that we have never yet understood, and for predicting aspects of their behaviour that we have never foreseen. But the models require a co-operative effort. As Marcel Boumans[6] claims for model-building in economics, knowledge must be collected from where we can find it, well outside the boundaries of what any single theory says, no matter how fundamental – and universal – we take that theory to be. And not just knowledge but guesses too. When we look at how fundamental theories get applied, it is clear that the *Ansatz* plays a central role.[7]

The Ginzburg-Landau model of superconductivity, which is described by

[6] Boumans 1998.

[7] It should be noted that this is not merely a matter of the distinction between the logic of discovery and the logic of justification, for my claim is not just about where many of our most useful representative models come from but also about their finished form: these models are not models of any of the theories that contribute to their construction.

Towfic Shomar in his PhD Dissertation,[8] gives a nice example of both the importance of co-operation and of the role of the *Ansatz*. As Shomar stresses, this model, built upwards from the phenomena themselves, is still for a great many purposes both more useful for prediction and wider in scope than what for years stood as the fundamental and correct model, by Bardeen, Cooper and Schrieffer (BCS).[9] The situation is reflected in the description in a standard text by Orlando and Delin of the development followed up to the point at which the Ginzburg-Landau model is introduced.[10] As Orlando and Delin report, their text started with electrodynamics as the 'guiding principle' for the study of superconductivity; this led to the first and second London equations.[11] The guiding discipline at the second stage was quantum mechanics, resulting in a 'macroscopic model' in which the superconducting state is described by a quantum wave function. This led to an equation for the supercurrent uniting quantum mechanical concepts with the electrodynamic ones underlying the London equations. The supercurrent equation described flux quantization and properties of type-II superconductors and led to a description of the Josephson effect. The third stage introduced thermodynamics to get equilibrium properties. Finally, with the introduction of the Ginzburg-Landau model, Orlando and Delin were able to add considerations depending on 'the bi-directional coupling between thermodynamics and electrodynamics in a superconducting system'.[12]

This kind of creative and co-operative treatment is not unusual in physics, and the possibility of producing models that go beyond the principles of any of the theories involved in their construction is part of the reason that modern physics is so powerful. So, under the influence of examples like the Ginzburg-Landau model, I would no longer make my earlier points by urging that the laws of physics lie, as they inevitably will do when they must speak on their own. Rather, I would put the issue more positively by pointing out how powerful their voice can be when put to work in chorus with others.

The first point I want to urge in this chapter then is one about how far the knowledge contained in the theories of physics can go towards producing accurate predictive models when these theories are set to work co-operatively with what else we know or are willing to guess for the occasion. But I shall not go into this in great detail since it is aptly developed and defended in the volume coming out of the research on our Modelling Project.[13] My principal thesis is less optimistic. For I shall also argue that the way our theories get

[8] Shomar 1998.
[9] Bardeen, Cooper and Schrieffer 1957.
[10] Orlando and Delin 1990.
[11] See Suárez 1998 for a discussion of these.
[12] Orlando and Delin 1990, p. 508.
[13] Morrison and Morgan 1998.

applied – even when they co-operate – puts serious limits on what we can expect them to do. My chief example will be of the BCS theory of superconductivity, which has been one of the central examples in the LSE Modelling Project. Readers interested in a short exposition of the core of the argument about the limits of theory in physics can move directly to Section 6.

2 The 'received view'

'Good theory already contains all the resources necessary for the representation of the happenings in its prescribed domain.' I take this to be a doctrine of the 'received' syntactic view of theories, which takes a theory to be a set of axioms plus their deductive consequences. It is also a doctrine of many standard versions of the semantic view, which takes a theory to be a collection of models.

Consider first the syntactic view. C. G. Hempel and others of his generation taught that the axioms of the theory consist of internal principles, which show the relations among the theoretical concepts, and bridge principles. But Hempel assigned a different role to bridge principles than I do. For Hempel, bridge principles do not provide a way to make abstract terms concrete but rather a way to interpret the terms of theory, whose meanings are constrained but not fixed by the internal principles. Bridge principles, according to Hempel, interpret our theoretical concepts in terms of concepts of which we have an antecedent grasp. On the received view, if we want to see how specific kinds of systems in specific circumstances will behave, we should look to the theorems of the theory, theorems of the form, 'If the situation (*e.g.*, boundary or initial conditions) is X, Y happens'.

Imagine for example that we are interested in a simple well-known case – the motion of a small moving body subject to the gravitational attraction of a larger one. The theorems of classical mechanics will provide us with a description of how this body moves. We may not be able to tell which theorem we want, though, for the properties described in the theory do not match the vocabulary with which our system is presented. That is what the bridge principles are for. 'If the force on a moving body of mass m is GmM/r^2, then the body will move in an elliptical orbit $1/r = 1 + ce\cos\phi$ (where e is the eccentricity; c a constant)'. To establish the relevance of this theorem to our initial problem we need a bridge principle that tells us that the gravitational force between a large mass M and a small mass m is of size GmM/r^2. Otherwise the theory cannot predict an elliptical orbit for a planet.

The bridge principles are crucial; without them the theory cannot be put to use. We may know for example from Schrödinger's equation that a quantum system with some particular initial state Ψ_i and Hamiltonian $\mathcal{H} = -(\hbar^2/2m)\nabla^2 + V(r) + (ie\hbar/mc)A(r,t)\cdot\nabla$ will evolve into the state Ψ_f. But this is of no practical consequence till we know that Ψ_i is one of the excited stationary states for the electrons of an atom, $\mathcal{H}$ is the Hamiltonian representing the interaction with the electromagnetic field and from Ψ_f we can predict an exponentially decaying probability for the atom to remain in its excited state. The usefulness of theory is not the issue here, however. The point is that on the 'received view' the theorems of the theory are supposed to describe what happens in all those situations where the theory matters, and this is true whether or not we have bridge principles to make the predictions about what happens intelligible to us. On this view the only problem we face in applying the theory to a case we are concerned with is to figure out which theoretical description suits the starting conditions of the case.

Essentially the same is true for the conventional version of the semantic view as well. The theory is a set of models. To apply the theory to a given case we have to look through the models to find one where the initial conditions of the model match the initial conditions of the case. Again it helps to have the analogue of bridge principles. When we find a model with an atom in state Ψ_i subject to Hamiltonian $\mathcal{H}$ we may be at a loss to determine if this model fits our excited atom. But if the atoms in the models have additional properties – *e.g.*, they are in states labelled 'ground state', 'first excited state', 'second excited state', and so on – and if the models of the theory are constrained so that no atom has the property labelled 'first excited state' unless it also has a quantum state Ψ_i, then the task of finding a model that matches our atom will be far easier. I stress this matter of bridge principles because I want to make clear that when I urge, as I do here and in chapter 3, that the good theory need not contain the resources necessary to represent all the causes of the effects in its prescribed domain, I am not just pointing out that the representations may not be in a form that is of real use to us unless further information is supplied. Rather I want to deny that the kinds of highly successful theories that we most admire represent what happens, in however usable or unusable a form.

I subscribe neither to the 'received' syntactic view of theories nor to this version of the semantic account. For both are cases of the 'vending machine' view. The theory is a vending machine: you feed it input in certain prescribed forms for the desired output; it gurgitates for a while; then it drops out the sought-for representation, plonk, on the tray, fully formed, as Athena from the brain of Zeus. This image of the relation of theory to the models we use to represent the world is hard to fit with what we know of how science works. Producing a model of a new phenomenon like superconductivity is an incredibly difficult and creative activity. It is how Nobel prizes are won. On the vending machine view you can of course always create a new theory, but there are only two places for any kind of human activity in deploying existing theory to produce representations of what happens, let alone finding a place for genuine creativity. The first: eyeballing the phenomenon, measuring it up, trying to see what can be abstracted from it that has the right form and combination that the vending machine can take as input; second – since we cannot actually build the machine that just outputs what the theory should – we either do tedious deduction or clever approximation to get a facsimile of the output the vending machine would produce.

This is not, I think, an unfair caricature of the traditional syntactic/semantic view of theory. For the whole point of the tradition that generates these two views is the elimination of creativity – or whim – in the use of theory to treat the world. That was part of the concept of objectivity and warrant that this tradition embraced.[14] On this view of objectivity you get some very good evidence for your theory – a red shift or a Balmer series or a shift in the trend line for the business cycle – and then that evidence can go a very long way for you: it can carry all the way over to some new phenomenon that the theory is supposed to 'predict'.

In *The Scientific Image*[15] Bas van Fraassen asks: why are we justified in going beyond belief in the empirical content of theory to belief in the theory itself? It is interesting to note that van Fraassen does not restrict belief to the empirical claims we have established by observation or experiment but rather allows belief in the total empirical content. I take it the reason is that he wants to have all the benefits of scientific realism without whatever the cost is supposed to be of a realist commitment. And for the realist there is a function for belief in theory beyond belief in evidence. For it is the acceptability of the theory that warrants belief in the new phenomena that theory predicts. The question of transfer of warrant from the evidence to the predictions is a short one since it collapses to the question of transfer of warrant from the evidence to the theory. The collapse is justified because theory is a vending machine: for a given input the predictions are set when the machine is built.

I think that on any reasonable philosophical account of theories of anything like the kind we have reason to believe work in the world, there can be no such simple transfer of warrant. We are in need of a much more textured, and I am afraid much more laborious, account of when and to what degree we might bet on those claims that on the vending machine view are counted as 'the empirical content' or the deductive consequences of theory. The vending machine view is not true to the kind of effort that we know it takes in physics to get from theories to models that predict what reliably happens; and the hopes that it backs up for a shortcut to warranting a hypothesised model for a given case – just confirm theory and the models will be warranted

[14] See Daston and Galison 1992.
[15] Van Fraassen 1980.

automatically – is wildly fanciful.[16] For years we insisted theories have the form of vending machines because we wished for a way to mechanise the warrant for our predictions. But that is an argument in favour of reconstructing theories as vending machines only if we have independent reason to believe that this kind of mechanisation is possible. And I have not seen even a good start at showing this.

3 Customising the models that theory provides

The first step beyond the vending machine view are various accounts that take the deductive consequences of a single theory as the ideal for building representative models but allow for some improvements,[17] usually improvements that *customise* the general model produced by the theory to the special needs of the case at hand. These accounts recognise that a theory may be as good as we have got and yet still need, almost always, to be corrected if it is to provide accurate representations of behaviour in its domain. They nevertheless presuppose that good scientific theories already contain representations of the *regular behaviours* of systems in their domain even though the predicted behaviours will not for the most part be the behaviours that occur.

To bring together clearly the main reasons why I am not optimistic about the universality of mechanics – or any other theory we have in physics, or almost have, or are some way along the road to having, or could expect to have on the basis of our experiences so far – I shall go step-by-step through what I think is wrong with the customisation story. The problem set to us is to predict or account for some aspect of the behaviour of a real physical system, say the pendulum in the Museum of Science and Industry that illustrates the rotation of the earth by knocking over one-by-one a circle of pegs centred on the pendulum's axis. On any of a number of customisation accounts,[18] we begin with an idealised model in which the pendulum obeys Galileo's law. Supposing that this model does not give an accurate enough account of the motion of the Museum's pendulum for our purposes, we undertake to customise it. If the corrections required are *ad hoc* or are at odds with the theory – as I have observed to be the usual case in naturally occurring situations like this – a successful treatment, no matter how accurate

[16] In order to treat warrant more adequately in the face of these kinds of observations, Joel Smith suggests that conclusions carry their warrant with them so that we can survey it at the point of application to make the best informed judgements possible about the chances that the conclusion will obtain in the new circumstances where we envisage applying it. See Mitchell 1997 for a discussion.

[17] It is important to keep in mind that what is suggested are changes to the original models that often are inconsistent with the principles of the theory.

[18] For example, Ronald Giere's. See Giere 1988.

and precise its predictions are, will not speak for the universality of the theory. So we need not consider these kinds of corrections here.

Imagine then that we are in the nice situation where all the steps we take as we correct the model are motivated by the theory; and eventually we succeed in producing a model with the kind of accuracy we require. What will we have ended up with? On the assumption that Newton's theory is correct, we will have managed to produce a blueprint for a nomological machine, a machine that will, when repeatedly set running, generate trajectories satisfying to a high degree of approximation not, I suppose, Galileo's law for a very idealised pendulum, but some more complex law; and since, as we are assuming for the sake of argument, all the corrections are dictated by Newtonian theory given the circumstances surrounding the Museum's pendulum, we will *ipso facto* have a blueprint for a machine that generates trajectories satisfying the general Newtonian law, $\mathbf{F} = m\mathbf{a}$. (Of course, the original ideal model was also a blueprint for a machine generating the $\mathbf{F} = m\mathbf{a}$ law.)

Once we have conceived the idealised and the deidealised models as nomological machines, we can see immediately what is missing from the customisation account. In a nomological machine we need a number of components with fixed capacities arranged appropriately to give rise to regular behaviour. The interpretative models of the theory give the components and their arrangement: the mass-point bob, a constraint that keeps it swinging through a small angle along a single axis, the massive earth to exert a gravitational pull plus whatever additional factors must be added (or subtracted) to customise the model. But that is not enough. Crucially, nothing must significantly affect the outcome we are trying to derive except for factors whose overall contribution can be modelled by the theory. This means both that the factors can be represented by the theory and that they are factors for which the theory provides rules for what the net effect will be when they function in the way they do in the system conjointly with the factors already modelled.

This is why I say, in talking of the application of a model to a real situation, that *resemblance is a two-way street*. The situation must resemble the model in that the factors that appear in the model must represent features in the real situation (allowing whatever is our favoured view about what it is to 'represent appropriately'). But it must also be true that nothing too relevant occurs in the situation that cannot be put into the model. What is missing from the account so far, then, is something that we know matters enormously to the functioning of real machines that are very finely designed and tuned to yield very precise outputs – the shielding that I have stressed throughout this book. This has to do with the second aspect of resemblance: the situation must not have extraneous factors that we have not got into the model. Generally, for naturally occurring systems, when a high degree of precision is to be hoped for, this second kind of resemblance is seldom achieved. For the theories we know, their descriptive capacities give out.

Let us lay aside for now any worries about whether corrections need to be made that are unmotivated by the theory or are inconsistent with it, in order to focus on the question of how far the theory can stretch. In exact science we aim for theories where the consequences for a system's behaviour can be deduced, given the way we model it. But so far the kinds of concepts we have devised that allow this kind of deducibility are not ones that easily cover the kinds of causes we find naturally at work bringing about the behaviours we are interested in managing. That is, as I have been arguing, why the laws of our exact sciences must all be understood with implicit *ceteris paribus* clauses in front. As I shall argue in the rest of this chapter, our best and most powerful deductive sciences seem to support only a very limited kind of closure: so long as the only relevant factors at work are ones that can be appropriately modelled by the theory, the theory can produce exact and precise predictions. This is in itself an amazing and powerful achievement, for it allows us to engineer results that we can depend on. But it is a long distance from hope that all situations lend themselves to exact and precise prediction.

4 Bridge principles and interpretative models

I have made a point of mentioning bridge principles, which get little press nowadays, because they are of central importance both practically and philosophically. Practically, bridge principles are a first step in what I emphasise as a *sine qua non* of good theory – the use of theory to effect changes in the world. They also indicate the limitations we face in using any particular theory, for the bridge principles provide natural boundaries on the domain the theory can command. So they matter crucially to philosophical arguments about the relations between the disciplines and the universal applicability of our favoured theories. These are arguments I turn to in later sections.

I take the general lack of philosophic investigation nowadays of what bridge principles are and how they function in physics to be a reflection of two related attitudes that are common among philosophers of physics. The first is fascination with theory *per se*, with the details of the formulation and the exact structure of a heavily reconstructed abstract, primarily mathematical, object: *theory*. I say 'heavily reconstructed' because 'theory' in this sense is far removed from the techniques, assumptions, and various understandings that allow what is at most a shared core of equations, concepts, and stories to be employed by different physicists and different engineers in different ways to produce models that are of use in some way or another in manipulating the world. The second attitude is one about the world itself, one I remarked on in the Introduction, an attitude that we could call *Platonist* or *Pauline*: 'For now we see through a glass darkly, but then face to face. Now I know in part; but then shall I know even as also I am known.'[19] It would be wrong to say, as a first easy description might have it, that these philosophers are not interested in what the world is like. Rather they are interested in a world that is not our world, not the world of appearances but rather a purer, more orderly world, a world which is thought to be represented 'directly' by the theory's equations. But that is not the world that contemporary physics gives us reasons to believe in when physics is put to work to manage what happens in the world.

As I have argued in the earlier chapters in this book, physics needs bridge principles because a large number of its most important descriptive terms do not apply to the world directly; rather, they function as abstract terms. The *quantum Hamiltonian*, the classical *force function* and the *electromagnetic field vectors* are all abstract. Whenever they apply there is always some more concrete description that also applies and that constitutes what the abstract concept amounts to in the given case. *Mass, charge, acceleration, distance*, and the *quantum state* are not abstract. When a particle accelerates at 32 ft/sec^2, there is nothing further that constitutes the acceleration. Similarly, although it may be complicated to figure out what the quantum state of a given system is, there is nothing more about the system that *is what it is* for that system to have that state. In chapter 2 we saw some simple examples.

For a more complex illustration consider the Hall effect. We start with what in classical electromagnetic theory serves as a very concrete description:[20] a conductor carries a uniform current density $\mathbf{J}$ of electric charge nq with velocity $\mathbf{v}$ parallel to, say, the y axis in a material of conductivity σ. So $J_y = nqv_y$. Ohm's law tells us how to put this in the more abstract vocabulary of electric fields:

$$\mathbf{J} = \sigma \mathbf{E}$$

Now we know that there is an electric field parallel to the current. What happens when the conductor is placed in an external magnetic field $\mathbf{B}$ parallel to the z axis? An electric field appears across the conductor in the direction of the x axis and the magnetic field exerts a force $F = qv_yB_z$ on the moving charge carriers in the current.

From our general knowledge of what forces do, we can then predict that the force will tend to displace the moving carriers in the x direction. This gives rise to a non-uniform charge distribution, which licenses, in turn, a new

[19] King James Bible, I Corinthians, 13:12.

[20] I put the point this way to help us keep in mind the fact that whether a description is concrete relative to a set of other descriptions depends on their use, and especially on how applications of the descriptions are warranted relative to one another.

electric field description in the x direction. Eventually an equilibrium is reached: the force exerted by virtue of the electric field, $F = qE_x$, balances that from the magnetic field:

$$qE_x + qv_yB_z = 0$$

Substituting the expression of **J** above, we can express the new electric field as

$$E_x = -(J_y/nq)B_z$$

$$= -R_HJ_yB_z,$$

where R_H is called the 'Hall coefficient'.[21] Throughout this account the expressions for the electric field and for the forces are used as abstract descriptions: they piggy-back on the more concrete descriptions in terms of the conductivity, the electric charge density, the velocity and the external magnetic field.[22] The magnetic field description piggy-backs, in turn, on some more concrete description involving its 'source' and material circumstances. This more concrete description is typically omitted from the discussion, though, since it is not directly relevant to the Hall effect. The example shows clearly that abstract concepts are in no way second-class citizens relative to the more concrete concepts which they depend on; on the contrary, these are just the concepts we need in physics for systematic explanation and prediction. To label a description as 'abstract' is to say something about how it attaches to the world, not about its power to explain what happens there. This is why bridge principles are so important in physics.

Abstract terms apply to the world via bridge principles. I shall show in the following sections how this feature of central concepts in physics delimits what most successful theories in physics can do. Classical mechanics, for instance, has enormous empirical success; but not a classical mechanics reconstructed without its bridge principles. When I say 'success' here, I am talking from an empiricist point of view. Whatever else we require in addition, a successful theory in physics must at least account in a suitable way for an appropriate range of phenomena. It must make precise predictions that are borne out in experiment and in the natural world. So when we think of reconstructing some object we call 'theory'[23] and we ask questions about,

[21] For more examples see Ohanian 1988 and Jackson 1975.

[22] As Jordi Cat has explained, a similar abstract-concrete relation involving electromagnetic descriptions and mechanical models was held by James Clerk Maxwell. According to Maxwell, with the illustration (and explanation) provided by mechanical models of electromagnetic concepts and laws, the gulf is bridged between the abstract and the concrete; electromagnetic forces and energy exist and can be understood clearly only in concrete mechanical models. See Cat 1995a and forthcoming.

[23] Or, for structuralists, some node in a theory net.

e.g., whether some kinds of descriptions are abstract relative to other kinds of descriptions in this theory, the answers must be constrained by considerations about what makes for the empirical success of the theory. Once we call this reconstructed object 'quantum theory' or 'the BCS theory of superconductivity', it will be reasonably assumed that we can attribute to this object all the empirical successes usually acknowledged for these theories. What this requirement amounts to in different cases will get argued out in detail on a case-by-case basis. The point here is about bridge principles. In the successful uses of classical mechanics, force functions are not applied to situations that satisfy arbitrary descriptions but only to those situations that can be fitted to one of the standard interpretative models by bridge principles of the theory; so too for all of physics' abstract terms that I have seen in producing the predictions that give us confidence in the theory.

Recall the analogue of this issue for the semantic view, again with the most simple-minded version of classical mechanics to illustrate. Does the set of models that constitutes the theory look much as Ronald Giere and I both picture it: pendulums, springs, 'planetary' systems, and the like, situated in specific circumstances, each of which also has a force acting on it appropriate to the circumstances; that is, do the objects of every model of the theory have properties marked out in the 'interpretative' models and a value for the applied force as well? Or, are there models where objects have *simply* masses, forces and accelerations ascribed to them with no other properties in addition? I think there is a tendency to assume that to be a scientific realist about force demands the second. But that is a mistake, at least in so far as scientific realism asserts that the claims of the theory are true and that its terms refer, in whatever sense we take 'true' and 'refer' for other terms and claims.

The term 'geodesic' is abstract, as, I claim, are many central terms of theoretical physics: it never applies unless some more concrete description applies in some particular geometry, *e.g.* 'straight line' on a Euclidean plane or 'great circle' on a sphere. But this does not mean that we cannot be realists about geodesics. The same holds for questions of explanatory or predictive or causal power. The set of models that I focus on, where forces always piggy-back on one or another of a particular kind of more concrete description, will predict accelerations in accordance with the principle $\mathbf{F} = m\mathbf{a}$. Still there is nothing across the models that all objects with identical accelerations and masses have in common except that they are *subject to the same force*. Putting the bridge principles into the theory when we reconstruct it does not conflict with realism. And it does produce for us theories that are warranted by their empirical successes.

5 Remarks on how representative models represent

My focus in this chapter is on interpretative models. I have little to say about how representative models represent, except to urge a few cautions about thinking of representations too much on the analogy of structural isomorphism. Consider the case of the quantum Hamiltonian, which is the example I will be developing in detail. I think it is important to use some general notion of representation of the kind R.I.G. Hughes develops[24] and not to think of the models linked to Hamiltonians as picturing individually isolatable physical mechanisms, otherwise we will go wrong on a number of fronts. First we can easily confuse my claim that Hamiltonians are abstract descriptions needing the more concrete descriptions provided by interpretative models with a demand that the Hamiltonians be explained by citing some physical mechanisms supposed to give rise to them. The two claims are distinct. Throughout quantum theory we regularly find bridge principles that link Hamiltonians to models that do not describe physical mechanisms in this way. The Bloch Hamiltonian discussed below provides an illustration.

Second, it could dispose one to a mistaken reification of the separate terms which compose the Hamiltonians we use in modelling real systems. Occasionally these Hamiltonians are constructed from terms that represent separately what it might be reasonable to think of as distinct physical mechanisms – for instance, a kinetic energy term plus a term for a Coulomb interaction. But often the break into separable pieces is purely conceptual. Just as with other branches of physics and other mathematical sciences, quantum mechanics makes heavy use of the method of idealisation and deidealisation. The Bloch Hamiltonian for the behaviour of moving electrons in a perfect crystal again provides an illustration. The usual strategy for modelling in condensed matter physics is to divide the problem artificially into two parts: (a) the ideal fictitious perfect crystal in which the potential is purely periodic, for which the Bloch Hamiltonian is often appropriate when we want to study the effects on conduction electrons; and (b) the effects on the properties of a hypothetical perfect crystal of all deviations from perfect periodicity, treated as small perturbations. This kind of artificial breakdown of problems is typical wherever perturbation theory is deployed, but it is not tied to perturbation analyses. As we shall see, the BCS theory relies on earlier treatments by Bardeen and Pines of the screened Coulomb potential that separates the long wavelength and short wavelength terms from the Coulomb interactions because it is useful to think of the effects of the two kinds of terms separately. But this is purely a division of the terms in a mathematical

[24] Hughes 1998.

representation and does not match up with a separation of the causes into two distinct mechanisms.

Third, without a broader notion of representation than one based on some simple idea of picturing we should end up faulting some of our most powerful models for being unrealistic. Particularly striking here is the case of second quantization,[25] from which quantum field theory originates. In this case we model the field as a collection of harmonic oscillators in order to get Hamiltonians that give the correct structure to the allowed energies. But we are not thus committed to the existence of a set of objects behaving just like springs – though this is not ruled out either, as we can see with the case of the phonon field associated with the crystal lattice described below or the case of the electric dipole oscillator that I describe in chapter 9.

Last, we make it easy to overlook the fact that when we want to use physics to effect changes in the world we not only need ways to link the abstract descriptions from high theory to the more concrete descriptions of models; we also need ways to link the models to the world. This is a task that begins to fall outside the interests of theorists, to other areas of physics and engineering. Concomitantly it gets little attention by philosophers of science. We tend to try to make do with a loose notion of resemblance. I shall do this too. Models, I say, *resemble* the situations they represent. This at least underlines the fact that in order for a description to count as a correct representation of the causes, it is not enough that it predicts the right effects; independent ways of identifying the representation as correct are required. I realise that this is just to point to the problem, or to label it, rather than to say anything in solution to it. But I shall leave it at that in order to focus on the separate problem of how we use the interpretative models of our theories to justify the abstract descriptions we apply when we try to represent the world. I choose the quantum Hamiltonian as an example. In the next section we will look in detail at one specific model – the BCS model for superconductivity – to see how Hamiltonians are introduced there.

6 How far does quantum theory stretch?

The Bardeen-Cooper-Schrieffer model of superconductivity is a good case to look at if we want to understand the game rules for introducing quantum Hamiltonians. As Towfic Shomar argues in his PhD Dissertation,[26] if we are looking for a quantum description that gives very accurate predictions about superconducting phenomena we can make do with the 'phenomenological' equations of the Ginzburg-Landau model. These equations are phenomenolo-

[25] Negele and Orland 1987.
[26] Shomar 1998.

gical in two senses: first, they are not derived by constructing a model to which a Hamiltonian is assigned, but rather are justified by an *ad hoc* combination of considerations from thermodynamics, electromagnetism and quantum mechanics itself. Second, the model does not give us any representation of the causal mechanisms that might be responsible for superconductivity. The first of these senses is my chief concern here.

The Ginzburg-Landau equations describe facts about the behaviour of the quantum state that, according to proper quantum theory, must be derived from a quantum Hamiltonian. Hence they impose constraints on the class of Hamiltonians that can be used to represent superconducting materials. But this is not the procedure I have described as the correct, principled way for arriving at Hamiltonians in quantum theory, and indeed the equations were widely faulted for being phenomenological, where it seems both senses of 'phenomenological' were intended at once. The description of the Ginzburg-Landau model in the recent text by Poole, Farachad and Creswick is typical: 'The approach begins by adopting certain simple assumptions that are later justified by their successful predictions of many properties of superconducting materials.'[27] Indeed it is often claimed that the Ginzburg-Landau model was not treated seriously until after we could see, thanks to the work by G'orkov, how it followed from the more principled treatment of the BCS theory.[28]

Before turning to the construction of the BCS Hamiltonian I begin with a review of my overall argument. We are invited to believe in the truth of our favourite explanatory theories because of their precision and their empirical successes. The BCS account of superconductivity must be a paradigmatic case. We build real operating finely-tuned superconducting devices using the Ginzburg-Landau equations. And, since the work of G'orkov, we know that the Ginzburg-Landau equations can be derived from quantum mechanics or quantum field theory using the BCS model. So every time a SQUID detects a magnetic fluctuation we have reason to believe in quantum theory.

But what is quantum theory? *Theory*, after all, is a reconstruction. In the usual case it includes 'principles' but not techniques, mathematical relations but little about the real materials from which we must build the superconducting devices that speak so strongly in its favour. *Theory*, as we generally reconstruct it, leaves out most of what we need to produce a genuine empirical prediction. Here I am concerned with the place of bridge principles in our reconstructed theories. The quantum Hamiltonian is abstract in the sense of 'abstract' I have been describing: we apply it to a situation only when that

[27] Poole, Farachad and Creswick 1995.

[28] *Cf.* Buckel 1991. It remains an interesting question whether this account is really historically accurate or rather reflects a preference in the authors for principled treatments.

situation is deemed to satisfy certain other more concrete descriptions. These are the descriptions provided by the interpretative models of quantum mechanics.

Albert Messiah's old text *Quantum Mechanics*[29] provides four basic interpretative models: the central potential, scattering, the Coulomb interaction and the harmonic oscillator – to which we should add the kinetic energy, which is taken for granted in his text. The quantum bridge principles give the corresponding Hamiltonians for each of the concrete interpretative models available in quantum mechanics. They provide an abstract Hamiltonian description for situations otherwise described more concretely. The point is: this is how Hamiltonians are assigned in a proper theoretical treatment; and in particular it is how they are assigned in just those derivations that we take to be the best cases where predictive success argues for the truth of quantum theory. When the Hamiltonians do not piggy-back on the specific concrete features of the model – that is, when there is no bridge principle that licenses their application to the situation described in the model – then their introduction is *ad hoc* and the power of the derived prediction to confirm the theory is much reduced.

The term 'bridge principle' is a familiar one. Like 'correspondence rule', 'bridge principle' has meant different things in different philosophical accounts. C. G. Hempel and Ernst Nagel worried about the meaning of theoretical terms. The internal principles can give them only a partial interpretation; bridge principles are needed to provide a full interpretation in a language whose terms are antecedently understood. Here I am not worried about questions of meaning, which beset all theoretical terms equally if they beset any at all. Rather, I am concerned about a distinction within theoretical concepts: some are abstract and some are not. Operationalists also use the terms 'bridge principle' and 'correspondence rule', but for them the bridge principles give rules for how to measure a quantity. Again this is not the use of 'bridge principle' I have in mind, for all quantities equally need procedures for how to measure them, whereas bridge principles, as I use the term, are needed only for the abstract terms of physics.

My claim about bridge principles and the limits of quantum physics is straightforward. Some theoretical treatments of empirical phenomena use *ad hoc* Hamiltonians. But these are not the nice cases that give us really good reasons to believe in the truth of the theory. For this we need Hamiltonians assigned in a principled way; and for quantum mechanics as it is practised that means ones that are licensed by principles of the theory – by bridge principles. In quantum theory there are a large number of derivative principles that we learn to use as basic, but the basic bridge principles themselves

[29] Messiah 1961.

are few in number. Just as with internal principles, so too with bridge principles: there are just a handful of them, and that is in keeping with the point of abstract theory as it is described by empiricists and rationalists alike.[30] We aim to cover as wide a range as we can with as few principles as possible.

How much then can our theories cover? More specifically, exactly what kinds of situations fall within the domain of quantum theory? The bridge principles will tell us. In so far as we are concerned with theories that are warranted by their empirical successes, the bridge principles of the theory will provide us with an explicit characterisation of its scope. The theory applies exactly as far as its interpretative models can stretch. Only those situations that are appropriately represented by the interpretative models fall within the scope of the theory. Sticking to Messiah's catalogue of interpretative models as an example, that means that quantum theory extends to all and only those situations that can be represented as composed of central potentials, scattering events, Coulomb interactions and harmonic oscillators.

So far I have mentioned four basic bridge principles from Messiah. We may expect more to be added as we move through the theory net from fundamental quantum theory to more specific theories for specific topics. Any good formalisation of the theory as it is practised at some specific time will settle the matter for itself. In the next section I want to back up my claim that this is how quantum theory works by looking at a case in detail. I shall use the Bardeen-Cooper-Schrieffer account of superconductivity as an example. This account stood for thirty-five years as the basic theory of superconductivity and, despite the fact that the phenomena of type-II superconductors and of high temperature superconductivity have now shown up problems with it, it has not yet been replaced by any other single account.

I chose this example because it was one I knew something about from my study of SQUIDs at Stanford and from our research project on Modelling at LSE. It turns out to be a startling confirmation of my point. The important derivations in the BCS paper are based on a 'reduced' Hamiltonian with just three terms: two for the energies of electrons moving in a distorted periodic potential and one for a very simple scattering interaction. This Hamiltonian is 'reduced' from a longer one that BCS introduce a page earlier. When we look carefully at this longer Hamiltonian, we discover that it too uses only the basic models I have already described plus just one that is new: the kinetic energy of moving particles, the harmonic oscillator, the Coulomb interaction, and scattering between electrons with states of well defined momentum, and then, in addition, the 'Bloch' Hamiltonian for particles in a periodic potential (itself closely related to the central potential, which is already among the

[30] For example, David Lewis, John Earman and Michael Friedman on the empiricist side, and on the rationalist, Abner Shimony.

basic models). Superconductivity is a quantum phenomenon precisely because superconducting materials[31] can be represented by the special models that quantum theory supplies. How much of the world altogether can be represented by these models is an open question. Not much, as the world presents itself, looks on the face of it like harmonic oscillators and Coulomb interactions between separated chunks of charge. Superconductivity is a case where, as we shall see, a highly successful representation can be constructed from just the models quantum theory has to offer. My point is that with each new case it is an empirical question whether these models, or models from some other theory, or no models from any theory at all will fit. Quantum theory will apply only to phenomena that these models can represent, and nothing in the theory, nor anything else we know about the structure of matter, tells us whether they can be forced to fit in a new case where they do not apparently do so.

7 Background on the BCS theory

In the BCS theory, as in earlier accounts of both ordinary conductors and of superconductors, the superconducting material is modelled as a periodic lattice of positive ions sitting in a sea of conduction electrons. Earlier theories by Werner Heisenberg, by Hans Koppe and by Max Born and K. C. Cheng had postulated Coulomb interactions between the conduction electrons to be responsible for superconductivity. But the discovery of the *isotope effect* simultaneously by Maxwell at the National Bureau of Standards and by Reynolds, Serin, Wright and Nesbitt at Rutgers, seemed to indicate that lattice vibrations play a central role. Experiments with different isotopes of the same material showed that the critical temperature at which superconductivity sets in and the critical value of a magnetic field that will drive a superconductor into the normal state depend strongly on the isotopic mass. So too do the vibrational frequencies of the lattice when the ions move like harmonic oscillators around their equilibrium positions (as they do in standard models where they are not fixed). Hence the hypothesis arises that lattice vibrations matter crucially to superconductivity. A first step towards the BCS theory came in earlier work by Hans Fröhlich and then by John Bardeen. Bardeen and Fröhlich separately showed that the potential due to electron interactions via lattice vibrations could be attractive, in contrast to the repulsive Coulomb potential between the electrons; and further that when the differences in energy between the initial and final states for the electrons interacting with the lattice were small enough, the overall effect would be attractive.

A second step was the idea of 'Cooper pairs'. These are pairs of electrons

[31] At least low temperature, type-I materials.

with well defined momenta which repeatedly interact via the lattice vibrations, changing their individual momenta as they do so but always maintaining a total null momentum. The Pauli exclusion principle dictates that no two electrons can occupy the same state. So normally at a temperature of absolute zero the lowest energy for the sea of conduction electrons is achieved when the energy levels are filled from the bottom up till all the electrons are exhausted. This top level is called the 'Fermi level'. So all the electron' energy will normally be below the Fermi level. The interaction of two electrons under an attractive potential decreases the total potential energy. Raising them above the Fermi level increases their energy of course. What Cooper showed was that for electrons of opposite momenta interacting through an attractive potential – Cooper pairs – the decrease in potential energy will be greater than the increase in kinetic energy for energies in a small band above the Fermi level. This suggested that there is a state of lower energy at absolute zero than the one in which all the levels in the Fermi sea are filled. This is essentially the superconducting state.

The first job of the BCS paper was to produce a Hamiltonian for which such a state will be the solution of lowest energy and to calculate the state. This is why the paper is such a good example for a study of how Hamiltonians get assigned. The construction of the Hamiltonian takes place in the opening sections of the paper. The bulk of the paper which follows is devoted to showing how the allowed solutions to the BCS Hamiltonian can account for the typical features of superconductivity. We will need to look only at the first two sections.

8 The BCS Hamiltonian

The BCS Hamiltonian, I have said, is a startling confirmation of my claim that the Hamiltonians we use to treat unfamiliar problems are the stock Hamiltonians associated with familiar models. The BCS Hamiltonian has four very familiar terms. The first two terms represent the energy of a fixed number of non-interacting particles – 'electrons' – with well-defined momenta. I put *electron* in scare quotes because the particles in the model are not ascribed all the properties and interactions that electrons should have. The right way to think about modelling in these kinds of cases seems to me to be to say that the particles in the model have just the properties used in the model. There is, nevertheless, a good reason for labelling them in certain ways – say, as *electrons* – because this suggests further features that if appropriately included, should lead to a better representative model, as both Mary Hesse and Imre Lakatos suggest. The third term in the BCS Hamiltonian represents the pairwise Coulomb interaction among the same electrons. The last term represents interactions that occur between pairs of electrons through

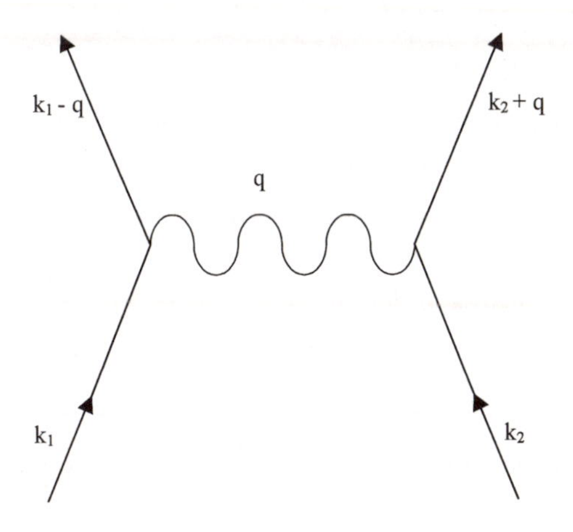

Figure 8.1 Electron-electron interaction through the lattice photons.
Source: Kittel 1963.

the exchange of a virtual phonon, as pictured in figure 8.1. This is a standard scattering interaction. The virtual phonons are associated with the lattice vibrations. ('Virtual' here means that energy conservation is briefly violated; phonons are the 'particles' associated with the energy field generated by vibrations of the lattice of ions.)

It may look from what I have said so far as if we just have stock models and Hamiltonians and that's it. Matters are not quite so simple as this initial description suggests, however. For it is not enough to know the form of the Hamiltonian; we have to know in some mathematical representation what the Hamiltonians actually are, and this requires more information about the models. Consider, for example, the first two terms in the BCS Hamiltonian. To specify what the two Hamiltonians are we will have to lay out the allowed values for the momenta of the electrons in the model; and to justify a given choice in these will require a lot more details to be filled in about the model. In fact exactly what structure the model actually has only becomes clear as we see how the third and fourth terms are developed since these latter two very simple terms appear at the cost of complication in the first two. What I shall describe is how each of these terms is justified.

The BCS Hamiltonian is not assigned in as principled a way as I may seem to be suggesting though. For a crucial assumption in the end is a restriction on which states will interact with each other in a significant way. The choice here is motivated by physical ideas that are a generalisation of those involved in Cooper's paper where electron pairs are introduced. BCS assume that scattering interactions are dramatically more significant for pairs of electrons

$$\mathcal{H} = \mathcal{H}_{\text{Bloch}} + \mathcal{H}_{\text{Coul.}} + \mathcal{H}_{\text{electron-phonon}}$$

$$\mathcal{H}_{\text{Bloch}} = \Sigma_{k>k_F} \varepsilon_{\mathbf{k}} n_{\mathbf{k}\sigma} + \Sigma_{k<k_F} |\varepsilon_{\mathbf{k}}|(1 - n_{\mathbf{k}\sigma})$$

$\mathcal{H}_{\text{Coul.}}$: Screened Coulomb Interactions

$$\mathcal{H}_{\text{electron-phonon}} = \tfrac{1}{2} \sum_{\mathbf{k},\mathbf{k}',\sigma,\sigma',\boldsymbol{\kappa}} \times \frac{2\hbar\omega_{\boldsymbol{\kappa}} |M_{\boldsymbol{\kappa}}|^2 c^*(\mathbf{k}' - \boldsymbol{\kappa}, \sigma') c(\mathbf{k}', \sigma') c^*(\mathbf{k} + \boldsymbol{\kappa}, \sigma) c(\mathbf{k}, \sigma)}{(\varepsilon_{\mathbf{k}} - \varepsilon_{\mathbf{k}+\boldsymbol{\kappa}})^2 - (\hbar\omega_{\boldsymbol{\kappa}})^2}$$

Figure 8.2 BCS-Hamiltonian. Source: Bardeen, Cooper and Schrieffer 1957.

with equal and opposite momenta. As in the earlier work of Fröhlich and Bardeen, electrons with kinetic energies in the range just above the Fermi level can have a lower total energy if the interaction between them is attractive. But there is a limit set on the total number of pairs that will appear in this range because not all pairs can be raised to these states since the Pauli exclusion principle prohibits more than one pair of electrons in a state with specific, oppositely directed values of the momentum. So here we see one of the features that quantum theory assigns to electrons that are retained by the electron-like particles in the model.

What is interesting for our topic is that these physical ideas are not built in as explicit features of the model that are then used in a principled way to put further restrictions on the Hamiltonian beyond those already imposed from the model. Rather the assumptions about what states will interact significantly are imposed as an *Ansatz*, motivated but not justified, to be tried out and ultimately judged by the success of the theory at accounting for the peculiar features associated with superconductivity.[32] Thus in the end the BCS Hamiltonian is a rich illustration: it is a Hamiltonian at once both theoretically principled and phenomenological or *ad hoc*. Let us look at each of the terms of the BCS Hamiltonian (figure 8.2) in turn.

Terms (1) and (2): Bloch electrons. The 'electrons' in our model are not 'free' electrons moving independently of each other, electrons moving unencumbered in space. They are, rather, 'Bloch' electrons and their Hamiltonian is the 'Bloch Hamiltonian'. It is composed of the sum of the energies for a

[32] Because of these *ad hoc* features it makes sense to talk both of the BCS theory and separately of the BCS model since the assumptions made in the theory go beyond what can be justified using acceptable quantum principles from the model that BCS offer to represent superconducting phenomena.

collection of non-interacting electrons. Each term in the sum is in turn a standard kinetic energy term for a moving particle – the most basic and well-known interpretative model in quantum theory – plus an externally fixed potential energy: $p^2/2m + V(r)$. In the Bloch Hamiltonian the second term for each electron represents the potential from the positive ions of the lattice, treated as fixed at their equilibrium positions. The crucial assumption is that this potential has the same periodicity as the lattice itself.

A second assumption fixes the allowed values for the energies that we sum over in the Bloch Hamiltonian. The energy is a function of the momentum, which, in turn, recalling wave-particle duality, is proportional to the reciprocal of the wave vector of the associated electron-wave. What facts in the model license a given choice? BCS adopt the Born-Karman boundary conditions on the superconducting material. The lattice is taken to be cubic (although the assumption is typically generalised to any parallelepiped) with the length on each side an integral multiple of the fixed distance between ions in the lattice. Well established principles from wave mechanics then dictate the allowed values for the wave vectors, hence the energies, in the Hamiltonian.

The German Structuralists[33] have taught us to think not of theories, but of theory-nets. The nodes of the net represent specialisations of the general theory with their own new principles, both internal principles and bridge principles. The introduction of the Bloch Hamiltonian in the quantum theory of conductivity is a good example of the development of a new bridge principle to deal with a specific subject. The Bloch Hamiltonian is an abstract description for situations that are modelled concretely as a certain kind of periodic lattice called a *Bravais lattice*: to be a moving electron in this kind of lattice is *what it is* for these electrons to be subject to a Bloch Hamiltonian. In the Bloch theory this term appears as a phenomenological Hamiltonian. It is not assigned by mapping out the details of the interaction between the (fictitious) non-interacting electrons and the ions that make up the (fictitious) perfect lattice, but rather represents the net effect of these interactions.

This is a good illustration of the difference between the phenomenological-explanatory distinction and the distinction between the principled and the *ad hoc* that I have already mentioned (although the term 'phenomenological' is often used loosely to refer to either or both of these). The search for explanation moves physicists to look for physical accounts of why certain kinds of situations have certain kinds of effects. Hamiltonians that pick out the putative physical mechanisms are called 'explanatory' as opposed to the 'phenomenological' Hamiltonians that merely produce quantum states with the right

[33] *Cf.* Stegmüller 1979; Gähde 1995.

kinds of features. The principled-*ad hoc* distinction, by contrast, depends on having an established bridge principle that links a given Hamiltonian with a specific model that licenses the use of that Hamiltonian. A Hamiltonian can be admissible under a model – and indeed under a model that gives good predictions – without being explanatory if the model itself does not purport to pick out basic explanatory mechanisms. This is just the case with the Bloch Hamiltonian for electrons moving in a Bravais lattice.

The treatment of Bloch electrons below the Fermi energy follows the same pattern.

Term (3): The screened Coulomb potential. Recall that a central claim of the BCS paper is that the Coulomb interactions between electrons should be less important to their quantum state than the interactions between them mediated by lattice vibrations, even though the Coulomb energies are much greater. Their argument depends on the fact that the Coulomb interactions are screened by the effects of the positive ions of the lattice. Electrons interacting under a Coulomb potential will repel each other and tend to move apart. But as they move apart the area in between becomes positively charged because of the ions which are relatively fixed there. So the electrons will tend to move back towards that area and hence move closer together. Bardeen and Pines[34] had shown that in their model the Coulomb interactions become effectively short-ranged, operating only across distances of the size of the inter-particle spacing, since the long-range effects can be represented in terms of high frequency plasma oscillations that are not normally excited.

The screened Coulomb potential is represented in the third term of the BCS Hamiltonian. To see where this term comes from consider a gas of electrons moving freely except for mutual interactions through Coulomb forces. The familiar bridge principle for the Coulomb interaction dictates that the contribution to the Hamiltonian for this gas due to the electron interactions should be $(1/2)\Sigma^{N}_{i=1}\Sigma^{N}_{j\neq i}\,(e^2/4\pi\varepsilon_0|\mathbf{r}_i - \mathbf{r}_j|)$ – that is, the usual Coulomb potential for a pair of electrons summed over all the ways of taking pairs. The model BCS want to study, though, is not of a gas of free interacting electrons but rather that of a gas of interacting electrons moving in a Bloch potential attributed to a lattice of fixed ions. As I have just described, for this model the Coulomb forces between electrons will be screened due to the effects of the fixed ions. BCS use the treatment of Bohm and Pines for this model, but they also mention one other established way to handle screening, namely, the Fermi-Thomas approach. This is the one Bardeen adopted to deal

[34] Bardeen and Pines 1955.

with screening in his 1956 survey.[35] For brevity I shall describe here only the Fermi-Thomas treatment.[36]

A usual first step for a number of approaches is to substitute a new model, the 'independent electron model', for the interacting Bloch electrons, a model in which the electrons are not really interacting but instead each electron moves independently in a modified form of the Bloch potential. What potential should this be? There is no bridge principle in basic theory to give us a Hamiltonian for this model. Rather the Hamiltonian for the independent electron model is chosen *ad hoc*. The task is to pick a potential that will give, to a sufficient degree of approximation in the problems of interest, the same results from the independent electron model as from the original model. We can proceed in a more principled way, though, if we are willing to study models that are more restrictive. That is the strategy of the Fermi-Thomas approach.

The Fermi-Thomas approach refines the independent electron model several steps further. First, its electrons are fairly (but not entirely) localized. These will be represented by a wave-packet whose width is of the order of $1/\boldsymbol{k}_F$, where $\boldsymbol{k}_F$ is the wave vector giving the momentum for electrons at the Fermi energy level. One consequence of this is that the Fermi-Thomas approach is consistent only with a choice of total potential that varies slowly in space. The Fermi-Thomas treatment also assumes that the total potential (when expressed in momentum space) is linearly related to the potential of the fixed ions. To these three constraints on the model a fourth can be consistently added, that the energy is modified from that for a free electron model by subtracting the total local potential. As a result of these four assumptions it is possible to back-calculate the form of the total potential for the model using standard techniques and principles. The result is that the usual Coulomb potential is attenuated by a factor $1/e^{k_F \cdot |\mathbf{r}_i - \mathbf{r}_j|}$. The Fermi-Thomas approach is a nice example of how we derive a new bridge principle for quantum theory. The trick here was to find a model with enough constraints in it that the correct Hamiltonian for the model could be derived from principles and techniques already in the theory. Thus we are able to admit a new bridge principle linking a model with its appropriate Hamiltonian. By the time of the BCS paper the Fermi-Thomas model had long been among the models

[35] Bardeen 1956.

[36] In fact I do this just for brevity but because the Bardeen and Pines approach introduces a different kind of model in which the electrons and ions are both treated as clouds of charge rather than treating the ions as an ordered solid. Tracing out the back and forth between this model and the other models employed which are literally inconsistent with it is a difficult task and the complications would not add much for the kinds of points I want to make here, although they do have a lot of lessons for thinking about modelling practices in physics.

available from quantum mechanics for representing conduction phenomena, and it continues to be heavily used today.

Term (4): Ion (or phonon) mediated interactions between electrons. So far we have considered models where the lattice ions have fixed positions. But, recall, we are aiming for a model where electrons interact through lattice vibrations. This kind of interaction was first studied by Fröhlich, following the Bloch theory of thermal vibrations and resistance in metals, where, Fröhlich observed, interactions between the lattice and the electrons involve the absorption and emission of non-virtual (i.e., energy conserving) vibrational phonons. For superconductors, he hypothesised, what matters is the absorption and emission of virtual phonons, as in figure 8.1. This is a very basic model and appears in almost identical form across elementary texts in quantum field theory as well as in texts at a variety of levels on superconductivity.

The first step in the construction of this term in the Hamiltonian is to represent the ions as a collection of coupled springs. Like the model of the Coulomb interaction, this is one of the fundamental models of quantum mechanics and the bridge rule that links it with the well-known *harmonic oscillator* Hamiltonian is one of the basic (i.e., non-derivative) principles of the theory. The periodic arrangement of the springs in the model fixes the allowed values of the frequency of oscillation. Like the electron-phonon exchange model, the spring model of ions is also pictured regularly in elementary texts in condensed matter physics and in superconductivity in an almost identical way. Figure 8.3 shows a typical example.

Although the harmonic oscillator model is basic in quantum mechanics, its use as a representation for the collection of ions in a metal is not simply *post hoc*. We do not use it just because it works but often justify it, instead, as an approximation from a prior model. In our case the prior model starts, as we have seen, with the picture of a lattice of metal atoms from which the valence electrons get detached and are able to move through the lattice, leaving behind positive ions which remain close to some equilibrium positions that fix the spacing of the lattice.

To get the harmonic model from the prior model, the chain of ions in the lattice are restricted to nearest-neighbour interactions. Thus the energy becomes a function of the total number of electrons and of the distance between any two neighbours. The harmonic model also assumes that ions move only a little about an equilibrium position – the displacements are small in comparison with the distances between ions. That means we can calculate the energy using a Taylor series expansion. Then we keep only the terms that in the expansion correspond to the energy at rest distance and to the second derivative of the energy with respect to the displacements from the distance

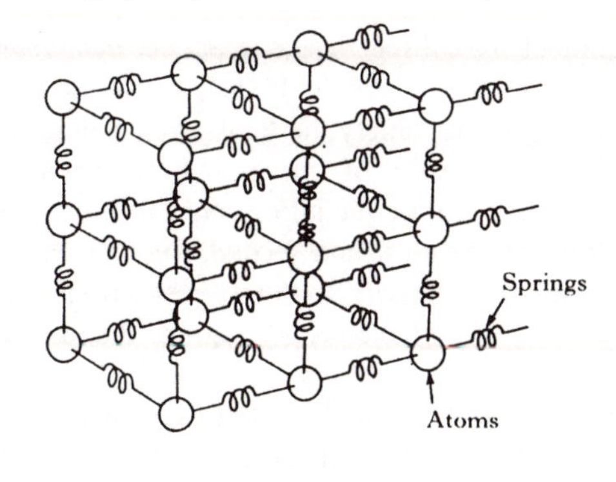

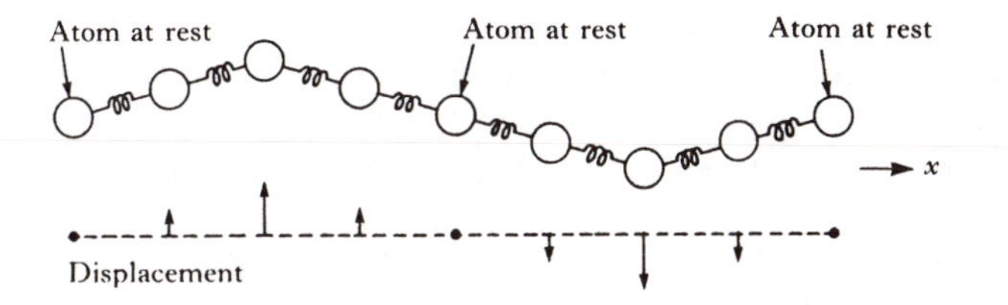

Figure 8.3 Spring model of solids. Source: Vidali 1993.

at rest.[37] The higher-order terms are the 'anharmonic' terms. In the harmonic approximation, for small values of the displacements, these are neglected.

The harmonic approximation is justified first of all on the grounds of its mathematical convenience. Harmonic equations of motion have exact solutions, whereas even the simplest anharmonic ones do not. Secondly, it allows us to get most of the important results with the simplest of models. Finally, it constitutes the basis for the perturbation method approach to the formalisation of complex cases. On this approach, more accurate results for the original model, which includes interactions between ions at separated sites, can be obtained by systematically adding (anharmonic) corrections to the harmonic term.

The underlying model that BCS work with is one in which there is a sea of loose electrons of well-defined momenta moving through a periodic lattice of positive ions whose natural behaviour, discounting interactions with the electron sea, is represented as a lattice of coupled harmonic oscillators, subject to Born-Karman boundary conditions. I shall call this the *full underlying model*. As I have already mentioned, a number of important features are

[37] The term corresponding to the linear expansion, i.e., the first derivative, is null, as it expresses the equilibrium condition for the distance at rest.

indicated about the model by calling the components in it 'electrons' and 'positive ions'. The two central ones are, first, the fact that the particles called 'electrons' are fermions, and thus satisfy the Pauli exclusion principle (hence it makes sense to talk, for instance, of filling each energy level up to the Fermi level since the exclusion principle allows only one electron per state), whereas the positive ions are bosons; and second, the fact that the particles in the model affect each other through Coulomb interactions. The Coulomb interactions occur pairwise among the electrons, among the ions, and between the electrons and the ions.

It is important to notice that this model is itself just a model. Even though I have called it 'the full underlying model', it is not the real thing, truncated. It is a *representation* of the structure of some given sample of superconducting material, not a literal *presentation*, as though seen from a distance so that only the dominant features responsible for superconductivity are visible. This more primary model is chosen with a look in two directions: to the theory on the one hand and to the real superconducting material on the other. The model aims to be explanatory. That means both that it should represent basic factors that will be taken as responsible for the superconductivity phenomena derived and that it must bring these factors under the umbrella of the theory, that is, it must use representations which we have a principled way of treating in the theory. It is significant for my central point about the limits of theory that the only kind of Hamiltonian used to describe this underlying model is the one for the Coulomb potential.

As I remarked in the discussion of term (3), it is impossible to solve the equation for the ground state of a Hamiltonian like this. So BCS substitute the new models I have been describing, with their corresponding Hamiltonians. Just as with term (3), if term (4) is to be appropriate, a lot more constraints must be placed on the underlying model before we can expect its results to agree with those from the BCS model. Similarly, the specific features of the Hamiltonians for the BCS model have to be constructed carefully to achieve similar enough results in the calculations of interest to those of the 'underlying' model.

The basis for the further term in the BCS model is the earlier work I mentioned by Hans Fröhlich. Fröhlich's model[38] begins with independent Bloch electrons with no interactions among them besides those which will appear as a consequence of the interaction between the electrons and the ions. In this model (as in Bloch's original one) the vibrations of the lattice are broken into longitudinal and transverse, and the electrons interact only with the longitudinal vibrations of the lattice. This, as Bardeen points out,[39] means

[38] Fröhlich 1950 and 1952.
[39] Bardeen 1956.

that the transverse waves are treated by a different model, a model in which the ions vibrate in a fixed negative sea of electrons. The Fröhlich model also assumes that the spatial vibration of the electron waves across the distance between ions in the lattice is approximately the same for all wave vectors. Using perturbation methods in the first instance and later a calculation that does not require so many constraints as perturbation analysis, Fröhlich was able to justify a Hamiltonian of the form of term (4), with an interaction coefficient in it reflecting the screening of the Coulomb force between ions by the motion of the electrons. The model assumes that in the presence of electron interactions the motion of the ions is still harmonic, with the frequencies shifted to take into account the screening of the ion interactions by the electrons. Similarly, in Fröhlich's version of term (4) the negative charge density in the model is no longer the density of the original Bloch electrons (which, recall, move in a fixed external potential) but rather that of electrons carrying information about the lattice deformation due to their interaction with it.

Now we can turn to the treatment by Bardeen and Pines. They have a model with plasma oscillations of the electron sea. They deal thus with a longer Hamiltonian involving energies of the individual Bloch electrons, the harmonic oscillations of the lattice, the harmonic oscillations of the plasma, an electron-lattice interaction, a plasma-phonon interaction, a plasma-electron interaction and a term for those wave vectors for which plasma oscillations were not introduced, including the shielded Coulomb electron-electron interaction. They rely on previous arguments of Pines' to show that this model can approximate in the right way a suitably constrained version of the 'full underlying model'. They are then able to show that the electron-lattice interaction can be replaced by a phonon-mediated electron-electron interaction described by a Hamiltonian of the form of term (4). In Bardeen and Pines' version of this term, as with Fröhlich's, the interaction coefficient is adjusted to provide approximately enough for the effects produced by shielding in the underlying model.

Other contemporary treatments. We have looked at the BCS theory in detail, but there is nothing peculiar about its use of quantum Hamiltonians. A number of accounts of superconductivity at the time treat the quantum mechanical superconducting state directly without attempting to write down a Hamiltonian for which this state will be a solution. This is the case, for instance, with Born and Cheng,[40] who argue from a discussion of the shape of the Fermi surface that in superconductivity spontaneous currents arise from a group of states of the electron gas for which the free energy is less with

[40] Born and Cheng 1948.

currents than without. It is also the case with Wentzel,[41] who treats superconductivity as a magnetic exchange effect in an electron fluid model, as well as Niessen,[42] who develops Heisenberg's account.

Those contemporary accounts that do provide Hamiltonians work in just the way we have seen with BCS: only Hamiltonians of stock forms are introduced and further details that need to be filled in to turn the stock form into a real Hamiltonian are connected in principled ways with features of the model offered to represent superconducting materials. By far the most common Hamiltonians are a combination of kinetic energy in a Bloch potential plus a Coulomb term. This should be no surprise since theorists were looking at the time for an account of the mechanism of superconductivity and Coulomb interactions are the most obvious omission from Bloch's theory, which gives a good representation of ordinary conduction and resistance phenomena. We can see this in Heisenberg's theory,[43] which postulates that Coulomb interactions cause electrons near the Fermi energy to form low density lattices where, as in Born and Cheng, there will be a lower energy with currents than without, as well as in Schachenmeier,[44] who develops Heisenberg's account and in Bohm,[45] who shows that neither the accounts of Heisenberg nor of Born and Cheng will work. Macke[46] also uses the kinetic energy plus Coulomb potential Hamiltonians, in a different treatment from that of Heisenberg.

Of course after the work of Bardeen and of Fröhlich studies were also based on the electron scattering Hamiltonian we have seen in term (4) of the BCS treatment, for instance in the work of Singwi[47] and Salam.[48] A very different kind of model studied both by Wentzel[49] and by Tomonaga[50] supposes that electron-ion interactions induce vibrations in the electron gas, where the vibrations are assigned the traditional harmonic oscillator Hamiltonian. Tomonaga, though, does not just start with what is essentially a spring model which he then describes with the harmonic oscillator Hamiltonian. Rather, he back-calculates constraints on the Hamiltonian from assumptions made about the density fields of the electron and ion gases in interaction. In these as in all other cases I have looked at in detail the general point I want to make is borne out. We do not keep inventing new Hamiltonians for each

[41] Wentzel 1951.
[42] Niessen 1950.
[43] Heisenberg 1947.
[44] Schachenmeier 1951.
[45] Bohm 1949.
[46] Macke 1950.
[47] Singwi 1952.
[48] Salam 1953.
[49] Wentzel 1951.
[50] Tomonaga 1950.

new phenomenon, as we might produce new quantum states. Rather the Hamiltonians function as abstract concepts introduced only in conjunction with an appropriate choice of a concrete description from among our set stock of interpretative models.

9 Stock models and the boundaries of the quantum theory

I have talked at length about *bridge principles* because our philosophical tradition is taken up with scientific knowledge. We focus on what we see as physics' claims about what properties things have and how they relate. Bridge principles have always been recognised as part of the body of precisely articulated knowledge claims. Our discussion reminds us that quantum physics provides more rules than these though for constructing theoretical representations.[51] In the case of the quantum Hamiltonian, bridge principles provide the form for the Hamiltonian from stock features that the model may have. Its detailed content is dictated by less articulated but equally well established techniques and constraints from other features of the model. The Born-Karman boundary conditions used to fix the allowed values of the momentum are an example. The point of my discussion here is that if we wish to represent a situation within quantum theory – within the very quantum theory that we prize for its empirical success – we must construct our models from the small stock of features that quantum theory can treat in a principled way. And this will fix the extent of our theory.

We are used to thinking of the domain of a theory in terms of a set of objects and a set of properties on those objects that the theory governs, wheresoever those objects and properties are deemed to appear. I have argued instead that the domain is determined by the set of stock models for which the theory provides principled mathematical descriptions. We may have all the confidence in the world in the predictions of our theory about situations to which our models clearly apply – like the carefully controlled laboratory experiments which we build to fit our models as closely as possible. But that

[51] It is well known that relativistic quantum field theory makes central the imposition of symmetry constraints, such as local gauge invariance, for determining the form of Lagrangians and Hamiltonians for situations involving interactions where the force fields are represented by gauge fields. But it cannot be claimed on these grounds alone that the Hamiltonians, say, are built, not from the bottom up, as I have discussed, but rather from the top down. Even in the case of interaction between electrons involving electromagnetic fields the local gauge symmetry requirement that establishes the form of the Lagrangian and Hamiltonian is only formal. It is satisfied by the introduction of a quantity that the symmetry requirement itself leaves uninterpreted. As Jordi Cat has argued, we still need in addition some bridge principles that tell us that the gauge field corresponds to electromagnetic fields and that indeed the Lagrangian or Hamiltonian describe a situation with some interaction with it (see Cat 1993 and 1995b). It is in this sense that it is often argued, mistakenly, that symmetry principles alone can dictate the existence of forces.

says nothing one way or another about how much of the world our stock models can represent.

ACKNOWLEDGEMENTS

This chapter is a shortened version, with some changes, of Cartwright 1998b. Thanks to both Jordi Cat and Sang Wook Yi for comments, for detailed discussion of both the physics and the philosophy and for help in the production. Research for the chapter was supported by the LSE Modelling in Physics and Economics Project, and a number of the ideas were developed during our group meetings.

9 How quantum and classical theories relate

1 Setting the problem

Consider a number of widely discussed topics in the philosophy of quantum mechanics: the measurement problem, the ever-expanding wave packet (Einstein's bed, which he feared would eventually fill his room), the possibility of state preparation, wave-particle duality. These problems are all reflections of the same basic phenomenon: for one and the same kind of system (say an electron) in order to describe adequately how it behaves in one context (if, for instance, we put a photographic plate in its path), we assign one quantum state to it at a particular time (in this case an eigenstate of position, or at least a state narrowly focused in position-space). But in order to represent the statistical results that would obtain in some different setting (imagine instead that we set a diffraction grating in its path followed by a detection screen) we assign a quite different state to it. The state in the latter case is generally a superposition of those states that describe the behaviour that might have been exhibited in the former setting. We can call this problem *the problem of superposition.*

There are broadly two strategies for solving the problem of superposition. The first is the strategy of *reduction.* When the system enters a setting in which its behaviour is correctly described by some component of the superposition, the state of the system physically changes from the superposition to one of the required components. This strategy faces three challenges.

(1) What are the states into which reduction occurs in a given reducing circumstance?

(2) In exactly which circumstances do reductions occur? This kind of change is not representable as a normal Schrödinger evolution. Hence on this proposal there are two distinct and different descriptions of how quantum states change. A realist understanding of the theory demands that there be some objective difference between the settings in which one description – the Schrödinger equation – applies and those in which the other – called 'reduction' – applies. Our drive to systematicity leads us (mistakenly, I think) to expect some simple universally applicable criterion to mark this difference.

What is this criterion? In the context of the measurement problem the assumption that reduction occurs on measurement raises the question, 'Exactly what is measurement?' The two most familiar answers are (a) an interaction with a system that has a very large number of degrees of freedom, and (b) an interaction with consciousness.

(3) Is the reduced state really the correct one? Once a criterion is proposed, a difficult job looms. Guarantees are required that the reduced state provides correct predictions for the future behaviour of systems in circumstances where reductions are supposed to occur. One standard way to attempt to achieve this is by trying to prove that the predictions from the reduced state (or the appropriately weighted mixture over the allowed reduced states) are not distinguishable from those from the original superposition. Presumably if there is a difference it is an empirical question which predictions are correct. So experimenting to find out is one of the tactics in defence of the reduction strategy.

The second major way of dealing with the problem of superposition is what I call the *have-your-cake-and-eat-it-too* strategy. This strategy maintains that in different ways both the superposition and the reduced states are correct. The modal interpretation, the Everett-Wheeler splitting universe, de Broglie's double solution, and various accounts using splitting minds as well as, I believe, Max Born's own understanding of his probabilistic interpretation are all versions of this strategy. In analogy with the reduction strategy, if we are to have our cake and eat it too we must be told (1) what states are allowed to the system in addition to the superposition and (2) in what circumstances the second states obtain. One favoured answer to the second question seems to be 'at the end of interaction'; and to the first question, that the allowed states are the states appearing in the bi-orthogonal decomposition of the superposition state representing the measured system plus measuring apparatus. That is, we take the superposition for the composite system and write it, as we always can, as a sum of products of states for the two interacting systems, where for each system the states that appear are mutually orthogonal. Then we say: *these* states are the states the interacting systems can have. This answer is popular despite the fact that it often gives physically counterintuitive results. Probably its popularity is based on two reasons. First, it seems to treat all systems identically so that no answer is required to the questions, 'Which interactions are measurements?' and 'Which systems are measuring apparatuses?' Second, so long as the original superposition is not degenerate, the states allowed by the bi-orthogonal rule are unique.[1]

The symmetry between the two interacting systems in the bi-orthogonal

[1] Robert Clifton gives a formal defence of the bi-orthogonal rule in Clifton 1995. For a discussion see Cartwright and Zouros in preparation.

decomposition is merely apparent, however. So long as the Hilbert spaces in which the two systems are represented are of the same dimension, all is well. But when they are of different dimensions, the system represented in the space of the lower dimension is privileged. In this case the bi-orthogonal rule will provide as possible states for the system represented in the lower dimensional space a set of basis states for that space. Whether this is physically reasonable or not, it is just what we have got used to from the case usually discussed, where the dimensions are equal. The states allowed to the other system, though, will not form a basis, but will be rather some (possibly very odd) superpositions over basis states, superpositions whose form is dictated by the insistence that the states of the first system form a complete set. The states for the second system then will still be orthogonal, but they will not be complete.

U. Fano in a paper in 1957[2] used this fact to show that the bi-orthogonal rule can be physically very implausible. In his example a collection of excited atoms make three transitions: $(l = 1, m_z = 0) \rightarrow (l' = 2, m'_z = 1)$; $(l = 1, m_z = 0) \rightarrow (l' = 2, m'_z = -1)$; and $(l = 1, m_z = 0) \rightarrow (l' = 2, m'_z = 0)$. The light given off in the first transition is left circular; in the second, right circular; and in the third, linearly polarised along the z axis. But that is not how it looks on the bi-orthogonal rule. For on this rule the two-dimensional photon space dominates. The photons are assigned states of opposite elliptical polarisation. The atoms in this case must be assigned different transitions from those we initially envisaged; not three as we first described but rather two, with the final states an unfamiliar superposition of those first supposed.[3]

Uniqueness, too, I find puzzling as a reason for choosing the bi-orthogonal rule since there are any number of rules that provide a unique assignment of states, and moreover, also give the same results as the bi-orthogonal rule in the case of simple von Neumann type-I measurements. I take it that this rule is thought somehow to be less arbitrary than any of the others. But why?

A third problem that besets the attempt to assign two states at once to quantum systems is that of formulating a book-keeping rule to tell which states are to be used for which purpose. Nor is it enough just to formulate the rule; we must also show both that the results of using the rule are always consistent with each other and that the rule provides empirically correct predictions. This is rarely done explicitly. Everett's rules for how the universe splits after interaction provide a kind of indirect book-keeping system for his account, and his analysis of what happens in the splitting universe when repeated measurements are made by different observers on the same system as well as his analysis of what happens when measurements are made not on

[2] Fano 1957.

[3] For a fuller discussion, see Cartwright 1974a and Del Seta and Cattaneo 1998.

a single system but on an ensemble of similarly prepared systems are part of the demonstrations that the results are consistent and correct. But ideally in this kind of an account we should like the rules and the demonstrations to be given an explicit formulation.

2 The classical-state alternative

I want to propose here a different kind of strategy for having your cake and eating it too, one with neither a universal rule for double state assigment nor a general demonstration of correctness and consistency. This is a strategy that instead undertakes to produce the requisite demonstrations in exactly those cases where detailed physical analysis results in two distinct representations for the same system.

The strategy employs a number of ideas that are taken from Willis Lamb. The ideas are developed in a series of papers of his on measurement.[4] I am going to locate these ideas of Lamb's in the frame I have just been sketching and then elaborate on some of the philosophical underpinning that supports the use I put them to. I will borrow two particular features of Lamb's account. The first is that the second set of states to be assigned beyond the conventional quantum superposition are not further quantum states but rather classical states. The second is that the assignment of these classical states is not a consequence of a general rule but comes instead from case-by-case analyses of specific physical situations. The view is, in a way I shall describe, peculiarly realistic about the quantum theory.

I begin with a standard assumption of the 'have-your-cake-and-eat-it-too' strategy: once a quantum state is assigned to a system in a particular kind of situation that can be represented with a quantum Hamiltonian, all future quantum states involving that system in that situation are determined by the Schrödinger equation in the ordinary way given its interactions. This is true no matter what other descriptions we assign to the system. This means that, so long as the system stays in situations where all the causes of change can be represented by a quantum Hamiltonian, once the system has been in an interaction it will never again have a state of its own but rather be represented as part of an ever increasing composite. (Of course there are familiar shortcuts for disregarding the rest of the composite in calculations where it is irrelevant.) The restriction to specific kinds of situations should be noted because Lamb does not think that quantum descriptions apply to systems in arbitrary situations. My agreement with Lamb on this is central to my view that quantum theory is severely limited in its scope of application. But that is not the central point in this

[4] Lamb 1969, 1986, 1987, 1989, 1993.

chapter, so for most of the discussion here the restriction to very particular kinds of situations will be repressed.

The core of the idea I want to develop here is that in addition to the quantum state, some systems will be correctly described by classical states as well. Both descriptions may be true of the system and true of it in exactly the same sense. Classical quantities (like the length of Einstein's bed or the position of particles on a photographic plate) depend on the classical state; and the quantum state fixes quantum mechanical features (like the dipole moment of the atoms in the semi-classical theory of the laser described below). There is thus no difficulty in knowing which states to use for which predictions. So, in this picture the book-keeping problem is trivial – at least at the formal level where these issues are generally discussed in the foundations of physics literature. I shall return to this issue, however, to explain why I add this caveat.

The idea that the states from the second set are classical is a simple one with obvious historical roots yet I do not think it has before been seriously considered in this setting. But classical states seems *prima facie* at least on a par with the more standard choice of quantum states with respect to both the book-keeping problem and to the question of when the second set of descriptions are supposed to apply. On other grounds classical states have a decided advantage over the quantum-state alternative:[5]

(a) Classical states have well-defined values for all classical observables. One of the chief reasons for wanting a second set of states is to deal with the problems of macroscopic objects that have interacted with quantum systems and might thereby become part of a composite superposition, objects like Schrödinger's cat. The second state is supposed to ensure that the macroscopic object nevertheless has well-defined values for the quantities we are used to seeing well-defined. But quantum states do not serve this purpose since systems with well-defined values for some observables will still be in superpositions across a variety of values of other observables that do not commute with the first. (b) Classical states do not spread as quantum states do. For example a system with a well-defined value of position will always retain a well-defined position under classical laws of evolution. So, Einstein's bed will not fill his room after all. (c) No rules for state selection are necessary. Problems like those I mentioned in discussing the bi-orthogonal rule do not arise. If we consider assigning some member of a second set of states to a system, the set of states is always the same – the set of classical states.

[5] These are also advantages that classical states have over the decoherence approach, which employs a form of the reduction strategy.

3 The relation between quantum and classical descriptions

Turn now to the second idea that contributes to the version of 'have-your-cake-and-eat-it-too' that I propose. The idea is that the assignment of classical states follows no formal universal rule. Rather it is part of the elaborate and ever expanding corpus of knowledge we acquire in quantum mechanics about how to solve problems and produce models for great varieties of different situations. This is a proposal that I have been long thinking about in my own way, so my remarks about it may not accurately reflect Lamb's view.

I begin with the familiar observation that the Schrödinger equation guarantees a deterministic evolution for the quantum state function. Nevertheless quantum mechanics is a probabilistic theory. What are the probabilities probabilities of? One answer is that they are probabilities that various 'observables' on the system possess certain allowed values. This is the answer associated with the *ensemble interpretation* of quantum mechanics. It is widely rejected on the grounds that results like those of the two-slit experiment show that observable quantities do not have 'possessed values' in quantum systems. The more favoured alternative is that the probabilities are probabilities for certain allowed values to be found when an observable is measured on a quantum system.

I reject both of these answers. I propose instead that the probabilities *when they exist at all* are probabilities for classical quantities to possess particular values. What is usually formalised as quantum theory is a theory about how systems with quantum states will evolve in the special circumstances that can be represented by quantum Hamiltonians. Of course our theoretical knowledge goes well beyond that, and we are often in a position (as I illustrate below) to use our quantum descriptions to help predict facts about the classical state of some system. As with Max Born's original treatment of scattering, these predictions may well be probabilistic, and it seems there is no way to eliminate the probabilistic element. But there is no guarantee that a quantum analysis will yield such predictions and there is no universal principle by which we infer classical claims from quantum descriptions. This is rather a piece-meal matter differing from case to case.

The fundamental picture I want to urge, following my interpretation of Lamb's ideas, is that in the right kind of situations some systems have quantum states, some have classical states, and some have both. The presupposition is that macro-objects in the usual settings that are well represented by classical models have classical states and that micro-objects in the kinds of settings that resemble quantum models have quantum states. But perhaps some macro-objects have quantum states as well. Macro-interference effects may give us one reason to assign a quantum state to macro-objects in certain kinds of situations. Sketches of micro-macro interactions like Schrödinger's cat or the von Neumann theory of measurement have also been thought to provide macro-quantum states. A third commonly accepted reason for assigning quantum states to macro-objects relies on standard procedures for modelling quantum interactions. It is supposed that (i) given any two systems with quantum states, the composite of the two must itself have a quantum state represented in the tensor product of the spaces representing the two separately; and also that (ii) all interactions between systems with quantum states are quantum interactions, representable by an interaction Hamiltonian on the tensor product space. Hence step by step the quantum states of a macro-object are built up out of the quantum states of the systems that make it up.

I myself do not find any of these reasons compelling, especially not the second and the third, which make claims about what kind of treatments quantum mechanics can in principle provide without actually producing the treatments. But that is not the most pressing matter. For the point is that there is no automatic incompatibility between quantum and classical states. Although contradictions may be unearthed in one case or another, they are not automatic.

What are classical states then? They are the states we represent in classical physics. Classically, the characteristics of physical systems are represented by analytic functions onto real numbers. In classical statistical mechanics, for instance, these quantities are functions of the basic dynamical variables, position and momentum, and, hence, they are defined on the n-dimensional phase space that the position and momentum values define. On this phase space physical states are represented by probability densities.

Position and momentum are also basic dynamical variables in quantum mechanics, although in this case they are represented mathematically by non-commuting Hermitian operators on a Hilbert space. It is on this complex vector space that quantum physical states are defined and can be associated with a statistical operator, the density matrix, that represents probability distributions. Despite some formal analogies, however, quantum and classical quantities fall under different mathematical characterisations, and there is no general rule that enables us to represent any classical quantity with a quantum one and *vice versa*. We do make attempts to formulate a general rule that enables us to represent classical quantities with quantum operators and *vice versa*, but they have all proven problematic.[6] In my opinion this is entirely

[6] Kochen and Specker argued in 1967 that the interpretive – or metaphysical – consistency of such a general mapping would require a hidden variable quantum theory, in which a phase space of hidden states has the formal structure of a classical phase space. However, they showed that no mapping exists that (1) allows the calculation of hidden variable theory expectation values by the classical method and (2) maps quantum operators onto real-valued functions on the space of hidden states in such a way that (2a) for any operator and

the wrong strategy. Though quantities represented in classical and quantum physics are mutually constraining (in different ways in different circumstances), they are different quantities exhibiting behaviour that is formalised differently in the two theories. We should not expect to be able to represent classical quantities in the quantum formalism, nor the reverse. In particular we should not expect to be successful in the familiar attempt to represent macroscopic classical quantities by a set of commuting operators in quantum theory.

There is one widely accepted assumption that offers a simple pattern of connection between quantum and classical states. That is the *generalised Born interpretation*. I think that the claim of this interpretation is unwarranted, though very specific instances of it may not be. (Born's own identification of $|\Psi(r)|^2$ with detection probabilities in scattering experiments is a good example of where the interpretation is warranted.) The generalised Born interpretation supposes that linear Hermitian operators

$$O = \Sigma\, e_i\, |\Phi_i\rangle\langle\Phi_i|$$

on a Hilbert space represent 'observable' quantities on the system represented by vector (Ψ) in that Hilbert space. The interpretation says that

$$\text{Prob}_\Psi\,(O = e_i) = \langle\Psi\, |\Phi_i\rangle\langle\Phi_i|\, \Psi\rangle$$

As before there are two views about the meaning of '$O = e_i$': (i) 'O possesses the value of e_i', or (ii) 'The value of e_i will be found in a measurement of O.'[7] Both readings have well-known problems. I think these problems arise because we are trying to find a simplistic route that does not exist for generating usable predictions and for testing the quantum theory. As is often remarked, the 'measurement' that appears in this interpretation is no real measurement, but something more like what a genie riding on the back of the particle would read without needing any instruments.[8] With quantum mechanics we can, and do, make predictions about what happens in real

its real counterpart, any function of the first also gets mapped onto the same function of the second, and (2b) preserves additivity relations for any two operators and their respective corresponding real counterparts. See Kochen and Specker 1967.

Mapping between classical and quantum quantities is also induced by the assignment of a joint probability distribution over the classical quantities. In the case of position and momentum the most well-known 'distribution' is Wigner's, which is associated with the rule

$$q^n\, p^m \rightarrow (1/2^n)\, \{\Sigma_{l=0}^{n}\, \tbinom{n}{l} Q^{n-l} P^m Q^l\}$$

But Wigner's distribution is not non-negative as a probability should be. In addition L. Cohen and M. Margenau have shown that in general no permissible distribution generates a rule for associating quantum operators with classical quantities that preserves algebraic relations. See Margenau and Cohen 1967 and Cartwright 1974b.

[7] Or, 'after some spontaneous reduction along the eigenstates of O.'

[8] *Cf.* James Park's measurement$_1$ and measurement$_2$ in Park 1969.

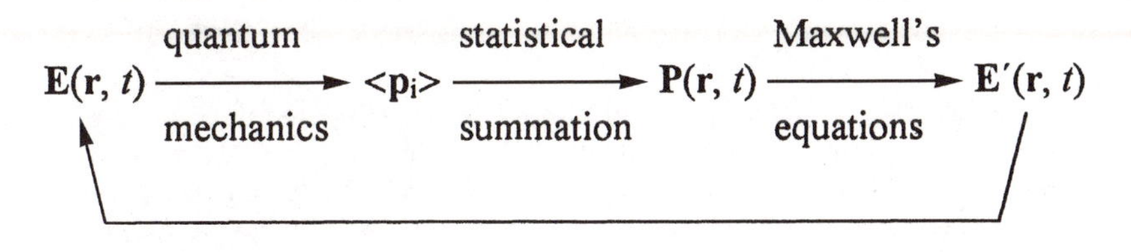

Figure 9.1 The semi-classical theory of the laser. Source: Sargent III, Scully and Lamb 1974.

measurement situations and in real devices such as spectroscopes and lasers and SQUIDs. But the analysis of all such situations is difficult, and the grounds for the inferences from the quantum analysis to predictions about the behaviour of the devices are highly various and context dependent.[9]

4 Some examples of quantum-classical connections

Lamb's own semi-classical theory of the laser provides a good example of classical-quantum relations. The theory assumes a classical electromagnetic field in interaction with the laser medium. The field induces a 'dipole moment' ($< \mathbf{p}_i > = < e\mathbf{r}_i >$) in the atoms of the medium; the dipole expectation is identified with the macroscopic polarisation of the medium; this in turn acts as a source in Maxwell's equations. Sargent, Scully and Lamb[10] diagram the process as in figure 9.1.

In the semi-classical theory the first step, linking the field and the dipole expectation, appears as a *causal interaction* between classical and quantum characteristics. In the step marked 'statistical summation' we see an *identification* of a quantum quantity, the suggestively named 'dipole expectation', with a purely classical polarisation. I put the term 'dipole expectation' in scare quotes to highlight the fact that from the point of view I am defending we should not think of it as a genuine expectation. I have rejected the assumption that every linear Hermitian operator represents some quantity (or, inaptly, some 'observable') which has or will yield values with probability distributions fixed by the generalised Born interpretation. Without the Born interpretation there are no automatic probabilities, and no expectations either.

The identification of $<e\mathbf{r}_i>$ with the macroscopic polarisation is a case in point. Here the 'interpretation' which proceeds by identifying these two is guided by a powerful heuristic: the quantum electric dipole oscillator. We are told by Sargent, Scully and Lamb, as we are in many texts on laser

[9] For a number of examples and a further discussion, see Chang 1997a, 1997b and also 1995.
[10] Sargent III, Scully and Lamb 1974, p. 96.

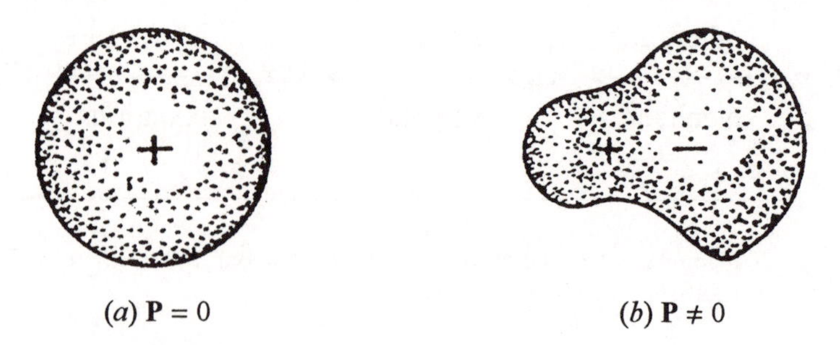

Figure 9.2 The quantum electric dipole oscillator. Source: Sargent III, Scully and Lamb 1974.

physics, that quantum electrons in atoms 'behave like charges subject to the forced, damped oscillation of an applied electromagnetic field'.[11] This charge distribution oscillates like a charge on a spring, as in Figure 9.2:

The discussions of the oscillator waver between the suggestive representation of $\langle e\mathbf{r}_i \rangle$ as a genuinely oscillating charge density and the more 'pure' interpretation in terms of time-evolving probability amplitudes. From the standard point of view this should be troubling; but without the generalised Born interpretation looming in the background I see nothing in this to worry about. The oscillator analogy provides a heuristic for identifying a quantum and a classical quantity in the laser model. The identification is supported by the success of the model in treating a variety of multimode laser phenomena – the time and tuning dependency of the intensity, frequency pulling, spiking, mode locking, saturation, and so forth.

Another example where a quantum characteristic is typically associated with a classical quantity is the standard semi-classical treatment of the Aharanov-Bohm effect. Consider Figure 9.3. In a two-slit experiment two diffracted electron beams produce an interference pattern on a screen. For the Aharanov-Bohm effect a solenoid is added at a required location between the slits and the screen. In that region a current in the solenoid produces an electromagnetic potential but no magnetic field. The electromagnetic vector potential is then *causally* associated with the occurrence of a phase shift in the interference fringes on the screen.

Let Ψ_0 be the wavefunction representing the electrons before the appearance of the vector potential. The wavefunctions corresponding to the paths 1 and 2 between slits 1 and 2, respectively, and an arbitrary point P on the screen are modified by the presence of the vector potential **A** in the following way:

[11] Sargent III, Scully and Lamb 1974, p. 31.

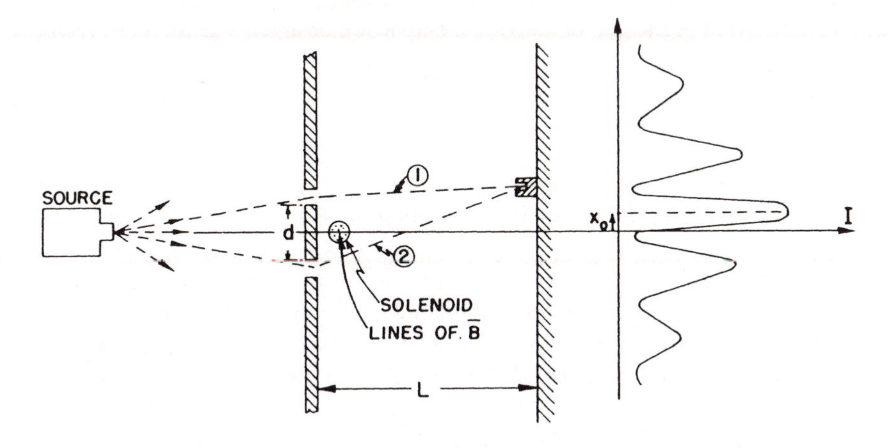

Figure 9.3 The Bohm-Aharanov effect. Source: Feynman, Leighton and Sands 1964.

$$\Psi_1 = \Psi_0 \, exp(-ie \, S_1/\hbar)$$

$$\Psi_2 = \Psi_0 \, exp(-ie \, S_2/\hbar)$$

where S_1 and S_2 are the action integrals, $S = \int \mathbf{A}d\mathbf{x}$, evaluated along the paths 1 and 2 respectively.

The phase shift at point P is

$$\Delta\delta = e(S_1 - S_2)/\hbar$$

which is directly proportional, by Stokes' theorem, to the magnetic flux Φ in the solenoid. Both the electromagnetic potential and the magnetic flux are classical quantities in this treatment, which facilitated the discovery of the Aharanov-Bohm quantum effect. This situation could even be perfectly conceived, untypically enough, as representing the use of a quantum characteristic – the phase shift – to measure a classical property – the magnetic flux.[12]

Both of these examples rely on semi-classical treatments, but there is nothing special about the semi-classical treatment. Exactly the same kind of identifications between quantum and classical expressions must be made at some point in any treatment that aims to get out predictions about classical quantities whether the field is taken to be quantised or not. Let us consider just one more case then, this time involving a quantised field. The 'fully' quantum theory of the laser, which we need in order to treat questions of line width, build-up from vacuum and photon statistics, will serve. Again, we can follow the treatment of Sargent, Scully and Lamb. At the core of this theory is the

[12] Thanks to Jordi Cat for suggesting this example.

identification of the magnitude of the classical multimode electric field vector

$$E(z,t) = \Sigma_s\, q_s(t)\, (2\Omega_s^2 m_s / V\varepsilon_0)^{1/2} \sin K_s z$$

with the quantum trace

$$E \sin Kz\ \mathrm{Tr}(\rho(t)(a + a^+))$$

where ρ is the density operator representing the field quantum mechanically. Again the identification is driven by a well-known heuristic, this time involving the harmonic oscillator analogy for changes in the electromagnetic field, which is used to introduce the annihilation and creation operators, a and a^+. Again the identification is borne out in the predictions it facilitates, such as the Lamb shift and the behaviour of lasers near threshold.

5 Avoiding problems on the quantum-classical approach

Consider now how problems are usually generated for views that assign a new state to the measuring device after measurement. There are two ways. First using von Neumann's simple theory of measurement it is supposed that the measurement interaction takes the composite $\Sigma c_i \Phi_i$ into the state $\Sigma c_i \Phi_i \Psi_i$ for some set of 'pointer position' eigenstates $\{\Psi_i\}$. When the interaction ceases each pointer must have a possessed value for position and thus be in one of the eigenstates Ψ_i. Since under the generalised Born interpretation pointer position i will occur with probability $|c_i|^2$, by observing the distribution of the pointer positions we can infer the values of the $|c_i|^2$. The problem arises when we consider the implications of the assumption that each pointer is in state Ψ_i for some i. In that case, it is argued, the state of the mated system in each case must be Φ_i since under the generalised Born interpretation the probability is zero for the pointer to state$_i$ and the system to have some different state$_j$ $(j \neq i)$ for the measured observable O. So for each individual composite the joint state must be $\Phi_i \Psi_i$; the ensemble of these composites will be then in the mixed state $\Sigma |c_i|^2 |\Phi_i\rangle\langle\Phi_i| \otimes |\Psi_i\rangle\langle\Psi_i|$. This contradicts the original assumption that the ensemble was in the pure, or non mixed, state $\Sigma c_i \Phi_i \Psi_i$.

This argument motivates advocates of the 'have-your-cake-and-eat-too' strategy to fix it so that the second set of states assigned (here $\{\Phi_i \Psi_i\}$) are true of systems in some different way than the original superposition so that the assignment of both the mixed and pure states will not be in contradiction. When the second set of states assigned is classical, no such assumption is yet necessary, for quantum and classical states do not contradict each other in this simple way. I say 'yet' because I take it that in each new analysis the question remains open. There are ways of inferring quantum states from information about classical states and the reverse, as we saw in Section 4.

Nothing I have said, or will say, guarantees that no situations will arise in which the classical information we can get in that situation and the inferences we take to be legitimate will produce a new quantum state for the apparatus and system that is inconsistent with the supposition that the Schrödinger-evolved superposition continues to be correct. But finding these situations, undertaking the analysis, conducting the experiments and drawing this surprising conclusion is a job that remains to be done. This is the job of *hunting the measurement problem*.

The second way to generate a measurement problem is to look at the time-evolved behaviour of the composite, system plus apparatus, to see if the predictions generated from the superposition contradict those from the second state assignment. As I remarked at the beginning, most versions of the 'have-your-cake-and-eat-too' strategy make some starting attempt to show that this can not happen. What happens in the version I propose? We imagine that after the measurement interaction the apparatus carries on with further interactions. Let us assume that the system it next interacts with has a quantum description. A problem could arise if this interaction also had a classical description so that the classical values assigned to the apparatus evolved as well. A contradiction is possible: we may find some interactions where the quantum-mechanically evolved state implies classical claims which contradict the information contained in the classically-evolved state. But such analyses are very uncommon and very difficult. We virtually never give a serious quantum-mechanical analysis of the continuing interactions of the measuring apparatus. To produce a contradiction we not only must do this but we must do this in a situation where we can draw classical inferences from the final quantum state as well. Again the job in front of us is not to solve the measurement problem *but to hunt it*.

6 The partnership between classical and quantum theory

How do we relate quantum and classical properties in modelling real systems? I maintain that in practice we follow no general formulae. The association is piecemeal and proceeds in thousands of different ways in thousands of different problems. Figuring out these connections is a good deal of what physics is about, though we may fail to notice this fact in our fascination with the abstract mathematical structures of quantum theory and of quantum field theory. Consider superconductivity. This is a quantum phenomenon. As we saw in the last chapter, we really do use quantum mechanics to understand it. Yet superconducting devices are firmly embedded in classical circuits studied by classical electronics. This is one of the things that most puzzled me in the laboratory studying SQUIDs. You would wire the device up, put it in the fridge to get it down below critical temperature, and then turn on

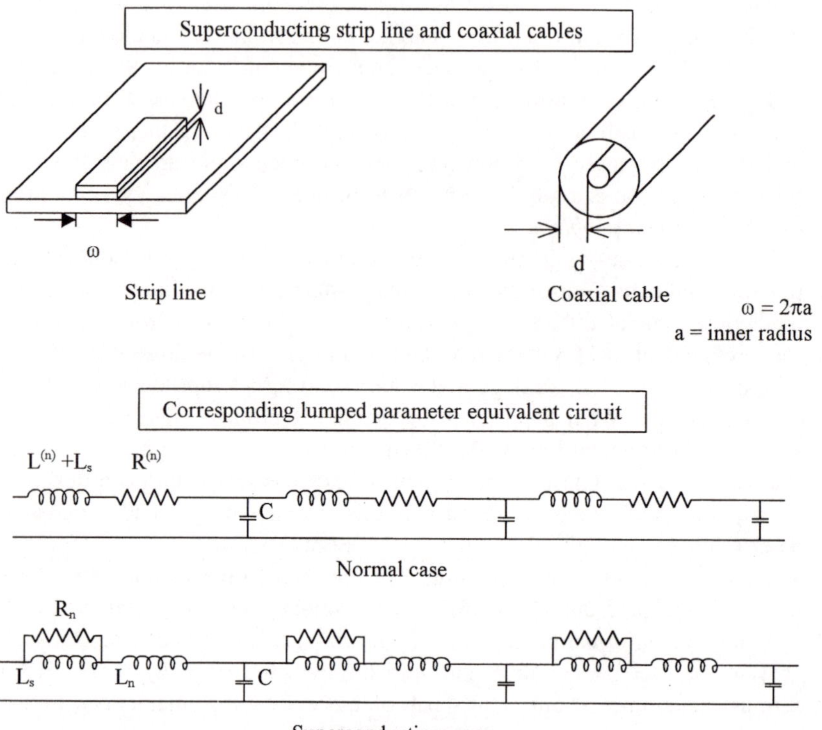

Figure 9.4 A Superconducting configuration and its corresponding classical circuit diagram. Source: Beasly 1990; recreation: George Zouros.

the switch. Very often you simply would not get the characteristic I-V curve which is the first test that the device is operating as a SQUID. What had gone wrong? To figure it out the experimenter would begin to draw classical circuit diagrams.

Without going all the way to SQUIDs we can see in figure 9.4 an example of the simplest kind of superconducting configuration that can be found in any standard treatment of superconducting electronics. 'What allows you to draw classical circuit diagrams for these quantum devices?' I would ask. The reply: 'There are well-known theorems that show that any complicated circuit is equivalent to certain simple circuits'. But that missed my point. What allows us to associate *classical* circuits with *quantum* devices?

No matter what theory you use – London, Ginzburg-Landau (G'orkov), BCS – all have as a central assumption the association of a familiar quantum quantity

Normal case	Superconducting case
$\delta^{-2} = \mu_0 \sigma^n \omega$ (skin depth)	$\lambda^{-2} = m/\mu_0 n e^s \omega$ (penetration depth)
with characteristic parameters	
L (per unit length) $= (\mu_0/\omega)(d + 2\delta)$	L (per unit length) $= (\mu_0/\omega)(d + 2\lambda)$
C (per unit length) $= \varepsilon\omega/d$	C (per unit length) $= \varepsilon\omega/d$
R^n (per unit length) $= (S/\delta\omega)$ $= 1/(\sigma^n \delta\omega)$	$R^{(s)}$ (per unit length) $= 2(\lambda/\delta)^3 R^{(n)}$
v (phase velocity) $= \sqrt{\frac{1}{LC}} = \bar{c}\sqrt{\frac{d}{d+2\delta}}$	$\nu = \bar{c}\sqrt{\frac{d}{d+2\lambda}}$
$\bar{c} = \sqrt{1/\mu_0\varepsilon}$	$\bar{c} = \sqrt{1/\mu_0\varepsilon}$
$z = \sqrt{\frac{L}{C}} = z_0\sqrt{\frac{d(d+2\delta)}{\omega^2}}$	$z = z_0\sqrt{\frac{d(d+2\lambda)}{\omega^2}}$

Figure 9.4 cont.

$$J_s = (e^*\hbar/2m^*i)\,(\Psi^*\nabla\Psi - \Psi\nabla\Psi^*) - (e^{*2}/m^*)|\Psi|^2\mathbf{A}^2 \quad ,$$

with a classical current that flows around the circuit. I say it is familiar because this is just what, in the Born interpretation, would be described as a probability current, taking $|\Psi|^2$ as a probability and using the conventional replacement for situations where magnetic fields play a role

$$\nabla \rightarrow \nabla \pm (ie^*/\hbar)\mathbf{A}$$

Yet we have all learned that we must not interpret $e|\Psi|^2$ as a charge density as Schrödinger wished to do. One of the reasons is that Ψ cannot usually be expressed in the co-ordinates of physical space but needs some higher dimensional co-ordinate space. But in this case it can be. And we have learned from the success of the theory that this way of calculating the electrical current is a good one.

But aren't we here seeing just the Born interpretation? No. On the Born interpretation what we have here is a probability and a probability in need of an elaborate story. The story must do one of two things. Either it both provides a mechanism that keeps reducing the paired electrons of the superconductor so that they are in an eigenstate of the current operator and also shows that this is consistent with the predictions that may well require a non-reduced state; or else in some way or another it ensures that the mean value (or something like it that evolves properly enough in time) is the one that almost always occurs. We have no such stories. More importantly, we need no such stories. This formula is not an instance of a general probabilistic interpretation of quantum mechanics but rather a special case of Schrödinger's identification between quantum and classical quantities. Although this identification is not acceptable as a general recipe, it works here. In this case we have, following Schrödinger, an empirically well-confirmed context-local rule for how a quantum state Ψ is associated with a classical current in a superconductor.

Let me give one more superconductivity example, this one from the treatment of Josephson junctions. There is one central way that we learn about the quantum state function that describes the electrons flowing through a Josephson junction. Whether we want to test the original claims about the systematic phase relations between the two sides of the junction, or about macroscopic quantum tunnelling, or about the limitations of the standard RSJ (relatively shunted junction) model, about Gaussian noise, shot noise, or various others, we use the same techniques. We produce I-V (current-voltage) curves like figure 9.5. The accompanying diagram makes a side point worth noting. One trick in designing an experiment is to figure out how different features of the quantum state will reveal themselves in modifications of the I-V curve. Another is to figure out how to ensure that what you are looking at is the I-V curve of the feature in the first place and not that of the rest of the circuitry in which it is embedded. After all, remember that for the most part superconductivity is a very low temperature phenomenon, so most of the junctions we study are stuck down at the bottom of a thermos about six feet tall.

The principal point I want to make about this example, though, is that we study the quantum state in Josephson junctions by looking at curves relating classical voltages and classical currents and not at all, as the Born interpretation suggests, by looking at probability distributions of allowed values for so-called quantum observables. We are looking in this case neither at a probability distribution nor at quantities represented by linear Hermitian operators. This is a case in which we use measurable quantities from other theories to probe the quantum state.

We can see the same point by looking at the converse process, in which we use the quantum state to calculate a quantity from classical circuitry theory. Let us consider an example that I looked at in studying tunnelling in Josephson junctions, the so-called 'normal resistance', R_N. This is the resistance of the junction when both metals are in the normal as opposed to the superconducting state – which they will be when the junction is not cooled to just a few degrees Kelvin:

$$R_N = (h^3/4\pi)/(e^2 N_r(0) \mathrm{N}_l(0) <|T^2|>)$$

where $N_r(0)$ and $N_l(0)$ are the densities of states at the Fermi level at the right

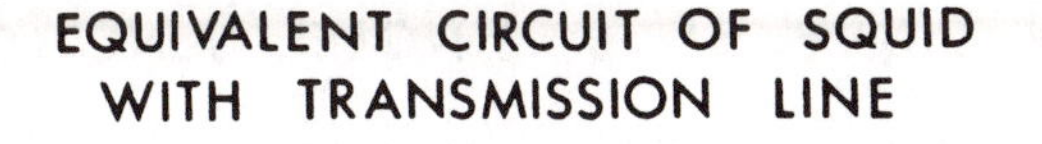

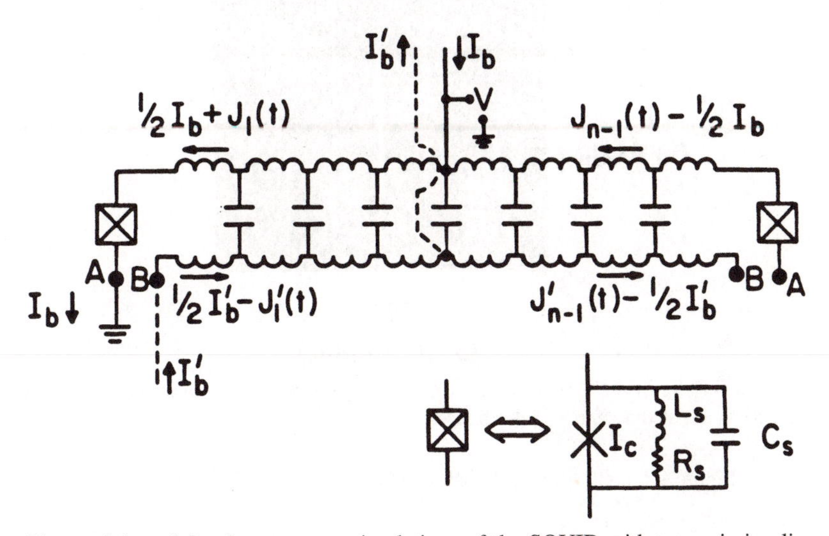

The model used for the computer simulations of the SQUID with transmission line. The RSJ has stray inductance Ls. The dimensionless capacitance of the transmission line is $B_{ctl} = 2\pi I_c R_S C_S/\phi_o$.

Figure 9.5 Typical equivalent current for a SQUID, with accompanying I–V curve.

and left metal respectively and $\langle|T^2|\rangle$ is the mean transition probability for an electron to move from the left to the right.

Where does this identification come from? I follow here the treatment of Antonio Barone and Gianfranco Paterno.[13] They first derive from fundamental microscopic theory an expression for the quasi-particle tunnelling current in a Josephson junction. (Quasi-particles are something like single electrons as opposed to the Cooper pairs which are responsible for superconductivity.)

$$I_{qp} = -\, h/eR_N \int d\omega\, n_l(\omega) n_r(\omega - eV_0/h)\ [f(\omega) - f(\omega - eV_0/h)]$$

Here $n_l(\omega)$ and $n_r(\omega)$ are the quasi-particle densities as a function of frequency on the left and on the right, $f(\omega) = 1/(e^{-\omega/k_B T} + 1)$ where k_B is the Boltzmann constant, and V_0 is the voltage across the junction. We now apply the formula

[13] Barone and Paterno 1982.

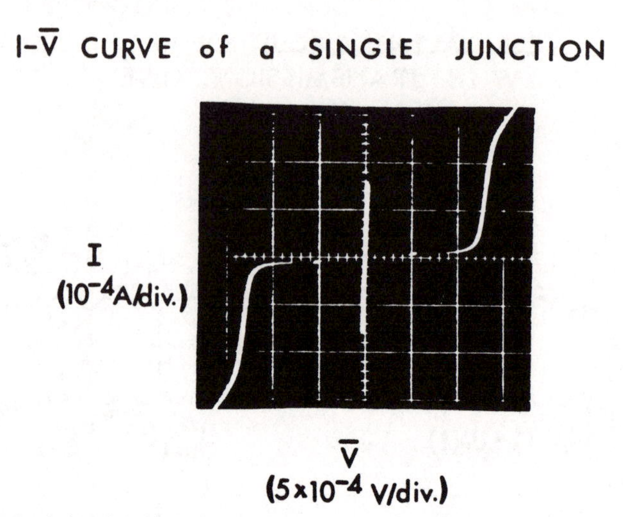

I–V characteristic of a single Josephson junction from the first generation. The zeros for I and V are at the centre of the figure.

Figure 9.5 cont. Source: Beasley 1990.

for I_{qp} to the case when both junction electrodes are in the normal state. Then $n(\omega) = 1$, and since

$$\int d\omega\, [f(\omega) - f(\omega - eV_0/h)] = eV_0/h$$

we get for the tunnelling current in the normal state (I_{NN})

$$I_{NN} = V_0/R_N$$

This shows that the quantity R_N defined above is the resistance of the junction when both metals are in the normal state.

There is one last point to make with this example. The identification of an expression we can calculate in quantum mechanics with the quantity R_N, which we can measure using Ohm's law with totally conventional volt-meters and ammeters, is not a once and for all identification. The expression for R_N depends very much on the specifics of the situation studied: here one simple quantum junction driven by a constant DC circuit with no spatial phase variation. In other situations we will need a very different quantum expression for the normal resistance. Finding it is just the kind of job that theoreticians working in superconductivity spend their time doing.

Examples like this show first that quantum mechanics and classical mechanics are both necessary and neither is sufficient; and secondly, that the two relate to each other differently from case to case. Both the conclusions speak in favour of using classical states if we propose to adopt a have-your-cake-and-eat-it too strategy for dealing with the superposition problem. On the other hand, they do discount the advantage I claimed for this approach with respect to the issue of book-keeping. For matters are not after all so simple as I first suggested. In one sense they are. Once we have a model which assigns states to the systems under study, book-keeping is straight forward on the classical-state view. Quantum features of the systems in the model are calculated from their quantum states and classical features from their classical states. But when we want to develop the model, in particular to think about how the systems might evolve, the underlying problems come to the fore once more.

To see how, consider the treatment of measurement in the modal interpretation. According to the modal interpretation, the observed outcome from a measurement interaction is to be accounted for by the 'second' state assigned to the measured system (the state 'true in the actual world' but not 'necessarily' true). The point of keeping the big superposition for the apparatus plus measured system is to account for the statistical features of any future measurement results on the composite in which interference would be revealed (*i.e.* to account for the wave aspect of the composite in its future behaviour). To account for the individual outcomes in the new measurement, though, a new set of second states comes into play. In early versions of the modal interpretation the exact nature of these second states and the probability distributions over them were to be fixed by the big evolved state for the composite in conjunction with information about which observables exactly were being measured in the second interaction. This simple version of the modal interpretation had little to say about the evolution of the second-level states: how does the state true of the measured system in the actual world at one time turn into the state true at another?

This is a question that both advocates and critics of the modal interpretation have put a good deal of effort into. It demands an answer just as much on the view I am defending, where the 'second' set of states are classical rather than quantum states. We might try the simple answer: classical states evolve by classical Langrangians into further classical states; quantum states evolve into quantum states under quantum Hamiltonians. But as I have indicated, this answer will not do from my own point of view. That is because of the possibility of mixed modelling that I have just been describing.

We can and do build models in which classical features affect the quantum description, and the reverse. Often the mixing is concealed at the very beginning or at the very end of the model. Quantum 'state preparation' is a clear general case of the first. You send your particles down a long tube in order to prepare a quantum state with a narrow spread in momentum or through a hole in a screen to prepare a localised wave packet. At the other end, we use the quantum state to calculate scattering amplitudes, or we use facts about the interference of wave functions in parallel arrays of Josephson junctions to calculate how the magnetic field applied to the array affects the total critical current. But there are a great many cases where the mixing occurs in the middle, or throughout, as in the self-consistent treatment of the laser described in section 4 and the calculation of normal resistance in particular kinds of SQUIDs, or the general account of superconductivity using a variety of different theories at different levels like that described by Orlando and Devlin, which we looked at in chapter 8. I am recalling these examples to stress the fact that the classical-state alternative does not provide an easy answer to the question about how the 'second' states evolve.

In fact, the perspective I advocate makes matters worse. For we also have no easy answer (even 'in principle') to the question of how the 'first' states evolve. The lesson is the same as the one I have been arguing throughout this book. Once we are entitled to represent a situation with a *pure* quantum model, whether from classical quantum mechanics or quantum electrodynamics or quantum field theory or some other special part of quantum theory, the appropriate theory will fix the evolution of whatever counts as the quantum state in that theory. But, as I have said repeatedly, we have no guarantees for any given situation that a pure quantum model will fit. This more general problem of state evolution is not readily apparent in the foundations of physics literature. I think that is because the literature tends to live inside the mathematical structure of quantum theory and seldom asks how that theory serves to model the world. Sometimes this is even done consciously, with the division of labour as a rationale: we study physics theory, modelling is a different matter, one of applied physics. My objection to this divide is that it is wrongly drawn. When we want to study how quantum mechanics treats measurements, then we need to look at quantum treatments of measurements, not rubrics for quantum treatments; just as when we want to study quantum treatments of superconductivity we need to read BCS and G'orkov and Anyon and not just look at our axiomatics. What is left out in the usual studies in foundations of physics is not how the physics is applied but rather, as Lamb urges,[14] the *physics*.

7 Philosophical remarks

I close with two brief philosophical discussions. First, what happens if we reject the generalised Born interpretation? How then do we interpret quantum mechanics? I should begin by pointing out that usually the discussion of the

[14] Lamb 1998.

measurement problem presupposes a strongly realistic view about the quantum state function. Instrumentalists who are prepared to use different incompatible state assignments in models treating different aspects of one and the same system are hardly troubled by the contradictions that are supposed to arise in special contexts of measurement. But it is puzzling why quantum realists should be calling for interpretation. For those who take the quantum state function seriously as providing a true and physically significant description, the quantum state should need no interpretation. There is no reason to suppose that those features of reality that are responsible for determining the behaviour of its microstructure must be tied neatly to our 'antecedent' concepts or to what we can tell by looking. Of course a fundamental property or state must be tied in some appropriate way to anything it explains. But laying out those ties need look nothing like an interpretation. Interpretations as standardly given in quantum mechanics attempt to exhaust what the quantum state is by reference to classical properties. The result is almost a reductive definition, except that the characterisation is not in terms of a set of possessed classical properties but rather in terms of dispositions to manifest these properties.

Behaviourism is an obvious analogy, and I think an instructive one. The distinction between behaviour and mental states is drawn in the first instance on epistemological grounds: it is supposed that we can know about or measure the behaviour of others in a way that we can never know about their mental states. Whatever is correct about behaviourism, the analogous distinction in quantum mechanics is a mistake. We know very well that we can measure quantum states, although, as I have argued, there is no universal template for how to go about it. So quantum mechanics is epistemologically in no worse position than classical mechanics. Clearly, for the behaviourist, more than epistemology is involved. The distinction takes on an important ontological status as well. Behaviour is thought to be the stuff really there; mental-state language merely a device for talking in a simple way about complicated patterns of behaviour. This is clearly not the kind of position that quantum realists want to find themselves suggesting about the quantum state.

My second remark is a brief one about the topic of co-operation that I broached at the beginning of the last chapter. Why do standard versions of the have-your-cake-and-eat-it-too strategy use quantum states rather than the more obvious choice of classical states that I urge, following Willis Lamb? I suspect the answer is that advocates of the strategy hold what I would take to be a mistaken view about what realism commits us to. Let us grant that quantum mechanics is a correct theory and that its state functions provide true descriptions. That does not imply that classical state ascriptions must be false. Both kinds of descriptions can be true at once and of the same system.

We have not learned in the course of our work that quantum mechanics is true and classical mechanics false. At most we have learned that some kinds of systems in some kinds of situations (*e.g.* electrons circulating in an atom) have states representable by quantum state functions, that these quantum states evolve and interact with each other in an orderly way (as depicted in the Schrödinger equation), and that they have an assortment of causal relations with other non-quantum states in the same or other systems. One might maintain even the stronger position that all systems in all situations have quantum states. As we have seen throughout this book I think this is a claim far beyond what even an ardent quantum devotee should feel comfortable in believing. But that is not the central matter here. Whether it is true or not, it has no bearing one way or another on whether some or even all systems have classical states as well, states whose behaviour is entirely adequately described by classical physics.

One reason internal to the theory that may lead us to think that quantum mechanics has replaced classical physics is reflected in the generalised Born interpretation. We tend to think that quantum mechanics itself is in some ways about classical quantities, in which case it seems reasonable for the realist to maintain that it rather than classical theory must provide the correct way of treating these quantities. But that is not to take quantum realism seriously. In developing quantum theory we have discovered new features of reality that realists should take to be accurately represented by the quantum state function. Quantum mechanics is firstly about how these features interact and evolve and secondly about what effects they may have on other properties not represented by quantum state functions.

In my opinion there is still room in this very realistic way of looking at the quantum theory for seeing a connection between quantum operators and classical quantities and that is through the correspondence rule that we use as heuristic in the construction of quantum mechanical Hamiltonians. But that is another long chapter. All that I want to stress here is my earlier remark that quantum realists should take the quantum state seriously as a genuine feature of reality and not take it is as an instrumentalist would, as a convenient way of summarising information about other kinds of properties. Nor should they insist that other descriptions cannot be assigned besides quantum descriptions. For that is to suppose not only that the theory is true but that it provides a complete description of everything of interest in reality. And that is not realism; it is imperialism.

But is there no problem in assigning two different kinds of descriptions to the same system and counting both true? Obviously there is not from anything like a Kantian perspective on the relations between reality and our descriptions of it. Nor is it a problem in principle even from the perspective of the naive realist who supposes that the different descriptions are in one-to-one correspondence with distinct properties. Problems are not there just because we assign more than one distinct property to the same system. If problems arise, they are generated by the assumptions we make about the relations among those properties: do these relations dictate behaviours that are somehow contradictory? The easiest way to ensure that no contradictions arise is to become a quantum imperialist and assume there are no properties of interest besides those studied by quantum mechanics. In that case classical descriptions, if they are to be true at all, must be reducible to (or supervene on) those of quantum mechanics. But this kind of wholesale imperialism and reductionism is far beyond anything the evidence warrants. We must face up to the hard job of practical science and continue to figure out what predictions quantum mechanics can make about classical behaviour.

Is there then a problem of superposition or not? In the account of quantum mechanics that I have been sketching the relations between quantum and classical properties is complicated and the answer is not clear. I do not know of any appropriately detailed analysis in which contradiction cannot be avoided. Lamb's view is more decided. So far we have failed to find a single contradiction, despite purported paradoxes like two-slit diffraction, Einstein Poldolsky-Rosen or Schrödinger's cat:

> There is endless talk about these paradoxes,
> Talk is cheap, but you never get more than you pay for.
> If such problems are treated properly by quantum mechanics, there are no paradoxes.[15]

Still many will find their suspicions aroused by these well-known attempts to formulate a paradox. In that case I urge that we look harder, and find the quantum measurement problem.

ACKNOWLEDGEMENTS

This chapter combines material previously published in Cartwright 1995b, 1995c and 1998a. Thanks to Jordi Cat for his help with the Bohm-Aharanov discussion and to Sang Wook Yi for comments.

[15] Lamb 1993.

Bibliography

Allport, P. 1991, 'Still Searching For the Holy Grail', *New Scientist* **5**, pp. 55–56.
1993, 'Are the Laws of Physics ''Economical with the Truth''?', *Synthèse* **94**, pp. 245–290.

Anand, S. and Kanbur, R. 1995, 'Public Policy and Basic Needs Provision: Intervention and Achievement in Sri Lanka', in Drèze, Sen and Hussain, 1995.

Anderson, P. W. 1972, 'More Is Different', *Science* **177**, pp. 393–396.

Andvig, J. 1988, 'From Macrodynamics to Macroeconomic Planning: A Basic Shift in Ragnar Frisch's Thinking', *European Economic Review* **32**, pp. 495–502.

Anscombe, E. 1971, *Causality and Determination*, Cambridge: Cambridge University Press. Reprinted in Sosa and Tooley 1993, pp. 88–104.

Aristotle [1970], *Aristotle's Physics: Book I and II*, trans. W. Charlton, Oxford: Clarendon Press.

Backhouse, R., Mäki, U., Salanti, A. and Hausman, D. (eds.) 1997, *Economics and Methodology*, International Economic Association Series, London: Macmillan and St Martin's Press.

Bacon, F. 1620 [1994], *Novum Organum*, Chicago: Open Court.

Bardeen, J. 1956, 'Theory of Superconductivity', in *Encyclopedia of Physics*, Berlin: Springer-Verlag, pp. 274–309.

Bardeen, J. and Pines, D. 1955, 'Electron-Phonon Interactions in Metals', *Physical Review* **99**, pp. 1140–1150.

Bardeen, J., Cooper, L. N. and Schrieffer, J. R. 1957, 'Theory of Superconductivity', *Physical Review* **108**, pp. 1175–1204.

Barone, A. and Paterno, G. 1982, *Physics and Applications of the Josephson Effect*, New York: Wiley.

Beasley, M. 1990, *Superconducting Electronics*, class notes (EE334), Stanford University.

Blaug, M. 1992, *The Methodology of Economics*, Cambridge: Cambridge University Press.

Bohm, D. 1949, 'Note on a Theorem of Bloch Concerning Possible Causes of Superconductivity', *Physical Review* **75**, pp. 502–504.

Born, M. and Cheng, K. C. 1948, 'Theory of Superconductivity', *Nature* **161**, pp. 968–969.

Boumans, M. 1999, 'Built-in Justification', in Morrison and Morgan 1999.

Brackenridge, J. B. 1995, *The Key to Newton's Dynamics*, Berkeley: University of California Press.

Brush, C. B. (ed.) 1972, *The Selected Works of Pierre Gassendi*, New York: Johnson Reprint Corporation.

Buchdahl, G. 1969, *Metaphysics and the Philosophy of Science*, Oxford: Blackwell.

Buckel, W. 1991, *Superconductivity: Foundations and Applications*, Weinheim: VCH.

Bunge, M. (ed.) 1967, *Quantum Theory and Reality*, Berlin: Springer-Verlag.

Cartwright, N. 1974a, 'A Dilemma for the Traditional Interpretation of Quantum Mixtures', in Cohen and Schaffner 1974.

1974b, 'Correlations without Joint Distributions in Quantum Mechanics', *Foundations of Physics* **4**, pp. 127–136.

1983, *How the Laws of Physics Lie*, Oxford: Oxford University Press.

1988, 'Capacities and Abstractions', in Kitcher and Salmon 1988.

1989, *Nature's Capacities and their Measurement*, Oxford: Clarendon Press.

1991, 'Fables and Models', *Proceedings of the Aristotelian Society*, Supplementary Volume **65**, pp. 55–68.

1992, 'Aristotelian Natures and the Modern Experimental Method', in Earman 1992.

1994, 'Fundamentalism vs. the Patchwork of Laws', *Proceedings of the Aristotelian Society* **103**, pp. 279–292.

1995a, '*Ceteris Paribus* Laws and Socio-Economic Machines', *The Monist* **78**, pp. 276–294.

1995b, 'Quantum Technology: Where to Look for the Quantum Measurement Problem', in Fellows 1995.

1995c, 'Where in the World is the Quantum Measurement Problem?', in Krüger and Falkenburg 1995.

1995d, 'Causal Structures in Econometric Models', in Little 1995.

1995e, 'Probabilities and Experiments', *Journal of Econometrics* **67**, pp. 47–59.

1997a, 'What is a Causal Structure', in McKim and Turner 1997.

1997b, 'Where Do Laws of Nature Come From?', *Dialectica* **51**, pp. 65–78.

1997c, 'Comment on ''Harold Hotelling and the Neoclassical Dream'', P. Mirowski and W. Hands', in Backhouse *et al.* 1997.

1997d, 'Models: Blueprints for Laws', *Philosophy of Science* **64**, pp. S292–S303.

1998a, 'How Theories Relate: Takeovers or Partnerships?', *Philosophia Naturalis* **35**, pp. 23–34.

1998b, 'Models and the Limits of Theory: Quantum Hamiltonians and the BCS Model of Superconductivity', in Morrison and Morgan 1999.

forthcoming (a), 'Causality, Independence and Determinism', in Gammermann forthcoming.

forthcoming (b), 'Causal Diversity and the Markov Condition', forthcoming in a special issue of *Synthèse* on causality, ed. Skyrms, B.

Cartwright, N., Cat, J., Fleck, L. and Uebel, T. E. 1996, *Otto Neurath: Philosophy Between Science and Politics*, Cambridge: Cambridge University Press.

Cartwright, N. and Zouros, G. in preparation, 'On Bi-orthogonal Decomposition'.

Cat, J. 1993, 'Philosophical Introduction to Gauge Symmetries', lectures, LSE, October 1993.

1995a, *Maxwell's Interpretation of Electric and Magnetic Potentials: the Methods of Illustration, Analogy and Scientific Metaphor*, PhD Dissertation, University of California, Davis.

1995b, 'Unity of Science at the end of the 20th Century: the Physicists Debate', MS, Harvard University Press.

1995c, 'The Popper-Neurath Debate and Neurath's Attack on Scientific Method', *Studies in the History and Philosophy Science* **26**, pp. 219–250.

1998, 'Unities of Science at the End of the 20th Century: A First Approximation to the Physicists Debate on Unification in Physics and Beyond', *Historical Studies in the Physical and Biological Sciences* **29**, pp. 2–46.

forthcoming, 'Understanding Potentials Concretely: Maxwell and the Cognitive Significance of the Gauge Freedom'.

Chang, H. 1995, 'The Quantum Counter-Revolution: Internal Conflicts in Scientific Change', *Studies in the History and Philosophy of Modern Physics* **26**, pp. 121–136.

1997a, 'On the Applicability of the Quantum Measurement Formalism', *Erkenntnis* **46**, pp. 143–163.

1997b, 'Can Planck's Constant be Measured with Classical Mechanics?', *International Studies in the Philosophy of Science* **11**, pp. 223–243.

Chiau, R. (ed.) 1996. *Amazing Light*, Berlin, Springer.

Clarke, J. and Koch, R. 1988, 'The Impact of High Temperature Superconductivity on SQUID Magnetometers', *Science* **242**, pp. 217–223.

Clifton, Rob 1995, 'Independently Motivating the Kochen-Dieks Modal Interpretation of Quantum Mechanics', *British Journal for Philosophy of Science* **46**, pp. 33–57.

Cohen, I. B. (ed.) 1958, *Isaac Newton's Papers and Letters*, Cambridge, MA: Harvard University Press.

Cohen, R. S. and Schaffner, K. F. (eds.) 1974, *PSA 1972: Proceedings of the 1972 biennial meeting, Philosophy of Science Association*, Dordrecht: Reidel.

Cohen, R. S., Hilpinen, R. and Renzong, Q. (eds.) 1996, *Realism and Anti-Realism in the Philosophy of Science*, Boston: Kluwer.

Collins, H. 1985, *Changing Order*, London: Sage Publications.

Dalla Chiara, M. L., Doets, K., Mundici, D. and van Benthem, J. (eds.) 1997, *Structures and Norms in Science*, Dordrecht: Kluwer.

Danto, W. and Morgenbesser, S. (eds.) 1960, *Philosophy of Science*, New York: New American Library.

Daston, L. and Galison, P. 1992, 'The Image of Objectivity', *Representations* **40**, pp. 81–128.

Davidson, D. 1995, 'Laws and Cause', *Dialectica* **49**, pp. 263–279.

Del Seta, M. and Cattaneo, G. 1998, 'Some Polarization Experiments and their Relevance for No-Collapse Interpretations of Quantum Mechanics', MS, LSE.

Ding, Z., Granger, C. and Engel, R. 1993, 'A long memory property of stock market returns and a new model', *Journal of Empirical Finance* **1**, pp. 83–106.

Drèze, J., Sen, A. and Hussain, A. (eds.) 1995, *The Political Economy of Hunger: Selected Essays*, Oxford: Oxford University Press.

Dupré, J. 1993, *The Disorder of Things*, Cambridge MA: Harvard University Press.

Earman, J. (ed.) 1992, *Inference, Explanation and Other Philosophical Frustrations*, Berkeley: University of California Press.

Edgeworth, F. Y. 1925, *Papers Relating to Political Economy*, 3 vols., London: Macmillan.

Everitt, C. W. F. (coordinator) 1980, *Report on a Program to Develop a Gyro Test of General Relativity*, Stanford CA: W. W. Hansen Laboratories, Stanford University.

Fano, U. 1957, ‘Description of States in Quantum Mechanics by Density Matrix and Operator Techniques’, *Review of Modern Physics* **29**, pp. 74ff.

Fellows, R. (ed.) 1995, *Philosophy and Technology*, Cambridge: Cambridge University Press.

Feynman, R. 1992, *The Character of Physical Law*, London: Penguin.

Feynman, R., Leighton, R. and Sands, M. 1964, *The Feynman Lectures on Physics*, Reading MA: Addison-Wesley.

Fine, A. 1986, *The Shaky Game*, Chicago: University of Chicago Press.

Foreman-Peck, J. 1983, *A History of the World Economy: International Economic Relations Since 1850*, Brighton: Wheatsheaf.

Fröhlich, H. 1950, ‘Theory of Superconducting State. I. The Ground State at the Absolute Zero of Temperature’, *Physical Review* **79**, pp. 845–856.

1952, ‘Interaction of Electrons with Lattice Vibrations’, *Proceedings of the Royal Society of London, Series A: Mathematical and Physical Science* **215**, pp. 291–298.

Gähde, U. 1995, ‘Holism and the Empirical Claim of Theory-Nets’, in Moulines 1995.

Galison, P. 1987, *How Experiments End*, Chicago: Chicago University Press.

1997, *Image and Logic*, Chicago: Chicago University Press.

forthcoming, ‘How Theories End’, in Wise forthcoming.

Gammermann, A. (ed.) forthcoming, *Causal Models and Intelligent Data Analysis*, New York: Springer-Verlag.

Gassendi, P. 1624 [1972], *Exercises Against the Aristotelians*, in Brush 1972.

Giere, R. 1984, *Understanding Scientific Reasoning*, Minneapolis: Minnesota University Press.

1988, *Explaining Science: A Cognitive Approach*, Chicago: University of Chicago Press.

1995, ‘The Skeptical Perspective: Science without Laws of Nature’, in Weinert 1995.

Glanvill, J. 1665 [1970], *Scepsis Scientifica*, Brighton: Harvester Press.

Glymour, C., Scheines, R. and Kelly, K. 1987, *Discovering Causal Structure: Artificial Intelligence, Philosophy of Science, and Statistical Modelling*, New York: Academic Press.

Goethe, J. W. 1810 [1967], *Theory of Colours*, trans. C. Eastlake, London: Cass.

1812 [1988], ‘The Experiment as Mediator between Object and Subject’, in Miller 1988.

Goldstein H. 1980, *Classical Mechanics*, Reading MA: Addison-Wesley.

Goodman, N. 1983, *Fact, Fiction, and Forecast*, Cambridge MA: Harvard University Press.

Haavelmo, T. 1944, ‘The Probabilistic Approach to Econometrics’, *Econometrica* **12**, *Supplement*, pp. 1–117.

Hacking, I. 1965, *Logic of Statistical Inference*, Cambridge: Cambridge University Press.

1983, *Representing and Intervening*, Cambridge: Cambridge University Press.

Hájek, A. 1997a, ‘A Probabilistic Magic Trick’, MS, California Institute of Technology.

1997b, ‘What Conditional Probability Could Not Be’, MS, California Institute of Technology.

Hands, D. W. and Mirowski, P. 1997, 'Harold Hotelling and the Neoclassical Dream', in Backhouse *et al.* 1997.

Harré, R. 1993, *Laws of Nature*, London: Duckworth.

Hart, O. and Moore, J. 1991, 'Theory of Debt Based on the Inalienability of Human Capital', *LSE Financial Markets Group Discussion Paper Series*, No. 129, London: LSE.

1994, 'Theory of Debt Based on the Inalienability of Human Capital', *Quarterly Journal of Economics* **109**, pp. 841–879.

Hausman, D. 1992, *The Inexact and Separate Science of Economics*, Cambridge: Cambridge University Press.

1997, 'Why Does Evidence Matter So Little To Economic Theory?', in Dalla Chiara, Doets, Mundich and van Beuttem, 1997, pp. 395–407.

Heisenberg, W. 1947, 'Zur Theorie der Supraleitung', *Zeitschrift für Naturforschung* **2a**, pp. 185–201.

Hempel, C. G. and Oppenheim, P. 1960, 'Problems in the Concept of General Law', in Danto and Morgenbesser 1960.

Hendry, D. F. 1995, *Dynamic Econometrics*, Oxford: Oxford University Press.

Hendry, D. F. and Morgan M. S. (eds.) 1995, *The Foundations of Econometric Analysis*, Cambridge: Cambridge University Press.

Hoover, K. forthcoming, *Causality in Macroeconomics*, Cambridge: Cambridge University Press.

Hotelling, H. 1932, 'Edgeworth's Taxation Paradox and the Nature of Demand and Supply Functions', *Journal of Political Economy* **40**, pp. 577–616.

Hughes, R. I. G. 1999, 'The Ising Model, Computer Simulation, and Universal Physics', in Morrison and Morgan 1999.

Hume, D. 1779 [1980], *Dialogues Concerning Natural Religion*, Indianapolis: Hackett.

Hutchison, T. W. 1938, *The Significance and Basic Postulates of Economic Theory*, New York: Kelley.

Jackson, J. D. 1975, *Classical Electromagnetism*, 2nd Edition, New York: Wiley.

Kitcher, P. and Salmon, W. (eds.) 1988, *Scientific Explanation*, Minneapolis: Univ. of Minnesota Press.

Kittel, C. 1963, *Quantum Theory of Solids*, New York: Wiley.

Kochen, S. and Specker, E. P. 1967, 'The Problem of Hidden Variables in Quantum Mechanics', *Journal of Mathematics and Mechanics* **17**, pp. 59–87.

Krüger, L. and Falkenburg, B. (eds.) 1995, *Physik, Philosophie und die Einheit der Wissenschaften*, Heidelberg: Spektrum Akademischer Verlag.

Kuhn, T. S. 1958, 'Newton's Optical Papers', in Cohen 1958.

Kyburg, H. 1969, *Probability Theory*, Englewood Cliffs NJ: Prentice Hall.

Lamb, W. 1969, 'An Operational Interpretation of Nonrelativistic Quantum Mechanics', *Physics Today* **22**, pp. 23–28.

1986, 'Quantum Theory of Measurement', *Annals of the New York Academy of Sciences* **48**, pp. 407–416.

1987, 'Theory of Quantum Mechanical Measurements', in Namiki *et al.* 1987.

1989, 'Classical Measurements on a Quantum Mechanical System', *Nuclear Physics B.* (Proc. Suppl.) **6**, pp. 197–201.

1993, 'Quantum Theory of Measurement: Three Lectures', lectures, LSE. July 1993.

1998, 'Superclassical Quantum Mechanics', MS, University of Arizona.
Lamb, W., Fearn, H. 1996, 'Classical Theory of Measurement', in Chiau 1996.
Lessing, G. E. 1759 [1967], *Abhandlungen über die Fabel*, Stuttgart: Philipp Reclam.
Lewis, D. 1986, *On the Plurality of Worlds*, Oxford: Blackwell.
Lindsay, R. 1961, *Physical Mechanics*, Princeton: D. Van Nostrand.
Little, D. (ed.) 1995, *On The Reliability of Economic Models*, Boston: Kluwer.
Macke, W. 1950, 'Über die Wechselwirkungen im Fermi-Gas', *Zeitschrift für Naturforschung* **5a**, pp. 192–208.
Mackie, J. L. 1965, 'Causes and Conditions', *American Philosophical Quarterly* **214**, pp. 245–264.
Mäki, U. 1996, 'Scientific Realism and Some Peculiarities of Economics', in Cohen *et al*. 1996.
Margenau, H. and Cohen, L. 1967, 'Probabilities in Quantum Mechanics', in Bunge 1967.
Maudlin, T. 1997, *A Modest Proposal on Laws, Counterfactuals and Explanation*, MS, Rutgers University.
McAllister, H. E. 1975, *Elements of Business and Economic Statistics*, New York: Wiley.
McGuinness, B. (ed.) 1987, *Unified Science*, Dordrecht: Reidel.
McKim, V. R. and Turner, S. P. (eds.) 1997, *Causality in Crisis?: Statistical Methods and the Search for Causal Knowledge in the Social Sciences*, Notre Dame: University of Notre Dame Press.
Mellor, H. 1995, *The Facts of Causation*, London: Routledge.
Menger, C. 1883 [1963], *Untersuchungen über die Methode der Sozialwissenschaften und der politischen Oekonomie insbesondere*, Leipzig: Duncker & Humblot 1883, trans. *Problems of Economics and Sociology*, Urbana: University of Illinois Press.
Messiah, A. 1961, *Quantum Mechanics*, Amsterdam: North-Holland.
Mill, J. S. 1836, 'On the Definition of Political Economy and the Method of Philosophical Investigation Proper to it', repr. in *Collected Works of John Stuart Mill*, vol. 4, Toronto: Toronto University Press.
1843, *A System of Logic*, repr. in *Collected Works of John Stuart Mill*, vols. 7–8, Toronto: Toronto University Press.
1872, 'On the Logic of the Moral Sciences', repr. in *Collected Works of John Stuart Mill*, vol. 8, Toronto: Toronto University Press.
Miller, D. (ed. and trans.) 1988, *Johann Wolfgang von Goethe: Scientific Studies*, New York: Suhrkamp.
Mitchell, S. 1997, 'Pragmatic Laws', *Philosophy of Science* **64**, pp. 468–479.
Morrison, M. 1997, 'Mediating Models: Between Physics and the Physical World', *Philosophia Naturalis* **34**.
forthcoming, *Unifying Theories: Physical Concepts and Mathematical Structures*. Cambridge: Cambridge University Press.
Morrison, M. and Morgan, M. (eds.) 1999, *Models as Mediators*, Cambridge: Cambridge University Press.
Moulines, C. U. 1995, *Structuralist Theory of Science: Focal Issues, New Results*, Berlin: de Gruyter.
Mulholland, H. and Jones, C. R. 1968, *Fundamentals of Statistics*, London: Butterworth.

Namiki, M. *et al.* (eds.) 1987, *Proceedings of the Second International Symposium on the Foundations of Quantum Mechanics: In the Light of New Technology*, Tokyo: Japan Physics Society.

Ne'eman, Y. 1986, 'The Problems in Quantum Foundations in the Light of Gauge Theories', *Foundations of Physics* **16**, pp. 361–377.

Negele, J. W. and Orland, H. 1987, *Quantum Many-Particle Systems*, Reading MA: Addison-Wesley.

Nemeth, E. 1981, *Otto Neurath und der Wiener Kreis: Wissenschaftlichkeit als revolutionärer politischer Anspruch*, New York: Campus.

Neurath, O. 1933, 'United Science and Psychology', in McGuinness 1987.

1935, 'Einheit der Wissenschaft als Aufgabe', *Erkenntnis* **5**, pp. 16–22, trans. 'The Unity of Science As Task' in Neurath 1983.

1983, *Philosophical Papers 1913–46*, ed. and trans. R.S. Cohen and M. Neurath, Dordrecht: Reidel.

Newton, I. 1729, *The Mathematical Principles of Natural Philosophy*, London: Motte.

1959–76, *The Correspondence of Isaac Newton*, ed. H. W. Turnbull *et al.*, 7 vols., Cambridge: Cambridge University Press.

Niessen, K. F. 1950, 'On One of Heisenberg's Hypotheses in the Theory of Specific Heat of Superconductors', *Physica* **16**, pp. 77–83.

Ohanian, H. C. 1988, *Classical Electromagnetism*, Boston: Allyn and Bacon.

Orlando, T. and Delin, K. 1990, *Foundations of Applied Superconductivity*, Reading MA: Addison-Wesley.

Park, J. 1969, 'Quantum Theoretical Concepts of Measurement', *Philosophy of Science* **35**, pp. 205–231.

Pearl, J. 1993, 'Comment: Graphical Models, Causality and Intervention', *Statistical Science* **8**, pp. 266–269.

1995, 'Causal Diagrams for Experimental Research', *Biometrica* **82**, pp. 669–710.

forthcoming, 'Graphs, Causality and Structural Equation Models', to appear in *Sociological Models and Research*, Special issue on causality.

Phillips, D. C. 1987, *Philosophy, Science and Social Inquiry*, Oxford: Permagon.

Pissarides, C. 1992, 'Loss of Skill During Unemployment and the Persistence of Unemployment Shocks', *Quarterly Journal of Economics* **107**, pp. 1371–1391.

Poole, C. P., Farachad Jr., H. A. and Creswick, R. J. 1995, *Superconductivity*, San Diego: Academic Press.

Robinson, E. A. 1985, *Probability Theory and Applications*, Boston: International Human Resources Development Corporation.

Ruben, D. H. 1985, *The Metaphysics of the Social World*, London: Routledge.

Runciman, W. G. (ed.) 1978, *Max Weber, Selections in Translation*, Cambridge: Cambridge University Press.

Russell, B. 1912–13, 'On the Notion of Cause', *Proceedings of the Aristotelian Society* **13**, pp. 1–26.

Ryle, G. 1949, *The Concept of Mind*, London: Barnes and Noble.

Salam, A. 1953, 'The Field Theory of Superconductivity', *Progress in Theoretical Physics* **9**, pp. 550–554.

Salmon, W. C. 1971 (with contributions by Jeffrey, R. C. and Greeno, J. G.), *Statistical Explanation and Statistical Relevance*, Pittsburgh: University of Pittsburgh Press.

Sargent III, M., Scully, M. O. and Lamb, W. Jr 1974, *Laser Physics*, Reading MA: Addison-Wesley.

Schachenmeier, R. 1951, 'Zur Quantentheorie der Supraleitung', *Zeitschrift für Physik* **129**, pp. 1–26.

Sen, A. 1981, 'Public Action and the Quality of Life in Developing Countries', *Oxford Bulletin of Economics and Statistics* **43**, pp. 287–319.

1988, 'Sri Lanka's Achievements: How and When', in Srnivasan and Bardham 1988.

Sepper, D. L. 1988, *Goethe contra Newton*, Cambridge: Cambridge University Press.

Shafer, G. 1996, *The Art of Causal Conjecture*. Cambridge MA: MIT Press.

1997, 'How to Think about Causality', MS, Faculty of Management, Rutgers University.

Shoemaker, S. 1984, *Identity, Cause and Mind: Philosophical Essays*, Cambridge: Cambridge University Press.

Shomar, T. 1998, *Phenomenological Realism, Superconductivity and Quantum Mechanics*, PhD Dissertation, University of London.

Singwi, K. S. 1952, 'Electron-Lattice Interaction and Superconductivity', *Physical Review* **87**, pp. 1044–1047.

Smart, J. J. C. 1963, *Philosophy and Scientific Realism*, London: Routledge.

Smith, C. and Wise, M. N. 1989, *Energy and Empire: A Biographical Study of Lord Kelvin*, Cambridge: Cambridge University Press.

Sosa, E. and Tooley, M. (eds.) 1993, *Causation*. Oxford: Oxford University Press.

Spirtes, P., Glymour C., Scheines, R. 1993, *Causation, Prediction and Search*, New York: Springer-Verlag.

Srnivasan, T. N. and Bardham, P. K. (eds.) 1988, *Rural Poverty in South East Asia*, New York: Columbia University Press.

Stegmüller, W. 1979, *The Structuralist View of Theories*, Berlin: Springer-Verlag.

Suárez, M. 1999, 'The Role of Models in the Application of Scientific Theories: Epistemological Implications', in Morrison and Morgan 1999.

Suppe, F. 1977, *The Structure of Scientific Theories*, Urbana: University of Illinois Press.

Suppes, P. 1984, *Probabilistic Metaphysics*, Oxford: Blackwell.

Toda, H. and Philips, P. 1993, 'Vector Autoregressions and Causality', *Econometrica* **61**, pp. 1367–93.

Tomonaga, S. 1950, 'Remarks on Bloch's Method of Sound Waves Applied to Many-Fermion Problems', *Progress in Theoretical Physics* **5**, pp. 544–69.

Van Fraassen, B. 1980, *The Scientific Image*, Oxford: Clarendon Press.

1989, *Laws and Symmetry*, Oxford: Clarendon Press.

1991, *Quantum Mechanics: An Empiricist View*, Oxford: Clarendon Press.

Vidali, G. 1993, *Superconductivity: The Next Revolution?*, Cambridge: Cambridge University Press.

Weber, M. 1951, *Gesammelte Aufsätze zur Wissenschaftslehre*, Tübingen: J. C. B. Mohr (Paul Siebeck).

Weinert, F. (ed.) 1995, *Laws of Nature: Essays on the Philosophical, Scientific and Historical Dimensions*, New York: Walter de Gruyer.

Wentzel, G. 1951, 'The Interaction of Lattice Vibrations with Electrons in a Metal', *Physical Review Letters* **84**, pp. 168–169.

Wise, N. (ed.) forthcoming, *Growing Explanations: Historical Reflections on the Sciences of Complexity*.

Index